SpringerBriefs in Molecular Science

SpringerBriefs in Molecular Science present concise summaries of cutting-edge research and practical applications across a wide spectrum of fields centered around chemistry. Featuring compact volumes of 50 to 125 pages, the series covers a range of content from professional to academic. Typical topics might include:

- A timely report of state-of-the-art analytical techniques
- A bridge between new research results, as published in journal articles, and a contextual literature review
- A snapshot of a hot or emerging topic
- An in-depth case study
- A presentation of core concepts that students must understand in order to make independent contributions

Briefs allow authors to present their ideas and readers to absorb them with minimal time investment. Briefs will be published as part of Springer's eBook collection, with millions of users worldwide. In addition, Briefs will be available for individual print and electronic purchase. Briefs are characterized by fast, global electronic dissemination, standard publishing contracts, easy-to-use manuscript preparation and formatting guidelines, and expedited production schedules. Both solicited and unsolicited manuscripts are considered for publication in this series.

Adi Wolfson

Chemodiversity and the Ecological Crisis

 Springer

Adi Wolfson
Sami Shamoon College of Engineering
Beer-Sheva, Israel

ISSN 2191-5407 ISSN 2191-5415 (electronic)
SpringerBriefs in Molecular Science
ISBN 978-3-032-07362-4 ISBN 978-3-032-07363-1 (eBook)
https://doi.org/10.1007/978-3-032-07363-1

This Springer imprint is published by the registered company Springer Nature Switzerland AG
The registered company address is: Gewerbestrasse 11, 6330 Cham, Switzerland

Preface

The world we live in is made up of innumerable particles: elementary particles, atoms and molecules, and a multitude of substances with different and diverse properties: natural and synthetic substances, organic and inorganic compounds, and pure and composite materials. All of these are found in outer space and on Earth, in the air, in water sources, and on land, as well as in plants ("flora") and animals ("fauna"). These particles and substances are constantly moving and changing their states and shapes, decomposing, and being created by means of a variety of physical processes and chemical reactions—as part of biogeochemical cycles, while absorbing and emitting energy in various forms. Furthermore, changes in matter and energy that occur in different times and places, alongside interactions between different substances and transitions between different forms of energy, sustain all the systems and processes in the Universe. This is equally true in the biotic environment (the "biosphere"), consisting of microorganisms, plants, animals, and the human body; in abiotic (lifeless) environments, such as the geosphere, the lithosphere, the hydrosphere, and the atmosphere; and in the entire ecosphere, comprising ecosystems composed of biotic and abiotic elements in complex interrelationships. Such ecosystems, large and small, for all their components, also interact in intricate mutual relationships. Moreover, this chemical diversity ("chemodiversity") indicates the richness of the spatial and temporal variations and changes—the amounts, types, and states of all the particles and substances in Nature—from elements and atoms, through molecules and compounds, to composite mixtures, solutions, and materials.

The Universe, which encompasses all physical space and time as we know it, contains "matter," defined as anything that has mass and occupies volume, or anything that we can "perceive" with our senses; and the Universe also contains "energy," which is "the ability or power to do work," meaning, "to perform an action or a process." The various substances and energies that constitute the Universe also define different environments and differentiate between them. Today, it is customary to distinguish between three main environments: the natural environment (the inanimate and the animate environments), the human environment, and the virtual environment.

Like the elements, the compounds in Nature also appear in different physical states (i.e., solid, liquid, gaseous, and plasma) and different types of aggregation, depending on their internal, atomic bonds and bonding between the atoms and molecules that compose them. Moreover, certain natural substances often appear as homogeneous or heterogeneous mixtures, consisting of two or more different molecules or compounds, mixed, but not linked by chemical bonds.

In general, it is possible to organize and sort the material resources in the Universe with the help of five primary hierarchies:

1. *The subatomic physical matter hierarchy*: ranging from elementary particles of matter and energy (such as quarks and leptons); to particles of energy (e.g., photons and bosons); to more complex subatomic particles (like protons, neutrons) and their antiparticles—which, based on current scientific knowledge, are indivisible.
2. *The chemical substance hierarchy*—chemical structures: from elements, appearing in Nature (as atoms or molecules); through molecules of organic and inorganic compounds (composed of different elements); to solutions and mixtures of several molecules, compounds, and composite materials.
3. *The geological hierarchy*—geological and geographical forms: from minerals and rocks, through soils and bodies of water, to landscape formations, such as mountains.
4. *The biomaterial hierarchy*—biological structures: genes that carry hereditary loads through living cells, consisting of different organelles that form the tissues and organs of living organisms.
5. *The ecological hierarchy*—ecological formations ranging from small single species (i.e., a certain species of a living creature) or a population of individuals of the same species found in a common habitat, to "biotic communities" or "biomes" (a biological society consisting of different populations sharing an ecosystem and its biogeochemical cycles), up to the "biosphere" (which represents all the ecosystems or habitable space in the world).

While substances are created and move all the time, and compounds are assembled and decomposed, if nuclear reactions are not involved, the basic elements of matter (i.e., sub-atomic particles and atoms) are preserved. The changes that each of these resources undergoes between the different stages in each hierarchy and the interactions between the different resources at the different levels alongside changes in place and time create a branched and complex network of processes, during which substances are constantly being created and broken down. The enormous variety of these substances represents the richness, abundance, and diversity of matter. The vast number of substances, the given amounts of each one and their many different types and properties—their chemodiversity—is the key to the continued existence of both the entirety of Nature and all of Humankind.

The processes that occur within ecosystems and between them are based on the transition of matter and energy between living and inanimate environments. For example, the movement of atoms or molecules during different biogeochemical cycles, as in the cyclic process of photosynthesis, during which inorganic compounds, such as water (H_2O) and carbon dioxide (CO_2), react in plants and algae in the presence of solar radiation to produce sugars, or when nutrients from the soil are taken up into plants, then eaten by animals, and finally returned to the soil when the animals die and decompose.

The accelerated growth of the human population over decades, along with the increase in the quality of life and burgeoning consumer culture, have been diverting more and more natural resources from the natural environment to the human environment for the benefit of human existence. Meanwhile, Humanity has also become mentally detached and physically distanced from Nature, interfering more and more with natural processes and, in parallel, producing a wide variety of synthetic compounds and developing man-made processes. In fact, it was recently found that the total mass of synthetic materials produced by Humanity so far—from buildings and infrastructures to mobile phones—is greater than the total "biomass" on Earth (i.e., larger than the total mass of all the plants, animals, fungi, and bacteria on the globe). Moreover, most man-made materials are not suitable for the natural environment and do not decompose naturally, and their production, use, and their ultimate waste treatments release many harmful pollutants into the environment. All the while, human activity has both short- and long-term effects, not only on local environment but also on global processes. Among the most dangerous negative effects has been Humanity's exceeding the natural, self-sustaining limits of Nature's ecosystems, harming the natural ecological supply and demand equilibrium (e.g., by destroying rainforests, polluting the oceans and the atmosphere, ruining natural habitats, killing off many species). The continued existence of balanced natural processes and protection of species and ecosystem diversity, while preserving of the global matter/energy equilibrium, are crucial for the survival of all life on Earth, since Humanity depends upon natural resources. Indeed, harming Nature threatens human welfare and survival.

The ongoing destructive changes in the type, quality, and quantity of natural resources caused by the addition of numerous man-made products to the natural environment have led to severe global crises: (1) the pollution or "chemical crisis," manifested by high amounts of a variety of toxic compounds that emitted into the air, and discharged into the soil and the water sources, causing severe damage to both the natural and human environments; (2) the "biodiversity crisis" or "biological crisis," defined as the depletion in the amount and type of species in Nature; and (3) the climate or "physical crisis," referring to frequent and extreme climatic changes: heat and cold waves, melting glaciers, rising sea levels, floods and mudslides, droughts and damages to agricultural crops, and more, happening on Earth these past few decades—especially as a result of man-made greenhouse gas (GHG) emissions that trap increased heat in the atmosphere (the "greenhouse effect"). As such, the global average temperatures on the Earth's surface has risen by over 1° from the beginning of Industrial Revolution and continues to climb ("global warming"). These three crises that intertwined and affect each other are three phases of the ecological crisis that threaten the natural systems and all of Humanity.

The ecological crisis is manifested by the loss of natural areas, overexploitation of natural resources, massive changes in species diversity in Nature, the ongoing emission of sewage, waste, and pollutants into the air, water sources, and soil, and by notable global warming and significant climate changes. All of these cause damage on the microscopic and macroscopic levels to the "ecosystem services" that supply water, food, fibers, and genes; "regulation services," like humidity and temperature control; and "support services," such as plant pollination, and the maintenance of soil fertility and groundwater quality; as well as causing serious changes, such as decreasing atmospheric ozone concentrations; constantly increasing atmospheric GHG concentrations; and lower concentrations of oxygen or acidification in the oceans and seas. The ecological crisis has far-reaching environmental and socio-economic effects.

Modern Humanity has changed the Earth and Nature, becoming an invading and dominating species, conquering many natural areas, depleting numerous natural resources, and producing countless synthetic and unknown materials, potentially harmful to Nature and/or incompatible with natural processes. By doing so, Humanity has caused irresponsible, and possible irreversible, damage to ecosystems, biological habitats, and biodiversity, even interfering in evolutionary processes—to the extent that this new era has been dubbed the "Age of Man," in which humans are changing and shaping Nature with their own hands, from the production of plastics and other synthetic materials yielding micro- and nanoparticulate pollutants in our atmosphere and oceans and on land, to emission of various radioactive elements, to the processes of genetic improvement and cloning. Such far-reaching changes are merely "symptoms of the disease" known as "the human crisis," short-sighted anthropocentric behavior, for personal benefit, irresponsibly overstepping the critical boundaries of the natural global systems. This human crisis manifests itself in daily human activity within economic systems that have not been adapted for the preservation of natural resources and processes and do not operate within the "planetary boundaries"; therefore, most routine human daily actions negatively affect Nature on Earth.

There is no doubt that we must change the current economic model, which encourages growth and consumption rather than prosperity. Moreover, this "Age of Man" and the significant and damaging impact of the ecological crisis it created on Nature and Humanity may also mark the beginning of a new great human revolution, the "ecological revolution" (over three centuries after the Industrial Revolution). Hopefully, this ecological revolution will be characterized by a notable change in human thought and social conduct, showing return to a balanced relationship between Humanity and the environment, alongside the safeguarding and renewal of vital natural resources.

Despite the great scientific knowledge amassed and the many advanced technologies that "progress" has brought to Humanity, the solutions to the great crises we are experiencing still require serious changes in perceptions, perspectives, and habits. Recognition and understanding that the human conduct causing the ecological crisis is immoral should lead us to associate what is right and proper with what is ultimately worthwhile, to strive for prosperity rather than growth, to reduce the wasteful consumption of the limited natural resources and prevent the emission of pollutants at the source. Enlightened Humanity must establish a new, global socio-economic order.

Beer-Sheva, Israel   Adi Wolfson

Competing Interests The author has no competing interests to declare that are relevant to the content of this manuscript.

Contents

About the Author

Adi Wolfson is a Hebrew poet, an expert in green chemistry, green engineering, and sustainability, and an avid eco-activist. He serves as a Professor of Chemical Engineering at the Shamoon College of Engineering (SCE) in Be'er-Sheva, Israel. Wolfson has published ten books in Hebrew (poetry, prose, and nonfiction), as well as four nonfiction works on sustainability published in English. He has authored numerous academic papers in the "green" fields of sustainable chemistry, engineering, and services. A founding member of SCE's Green Processes Center, Wolfson collaborates with national and local authorities and with Israeli industries toward the promotion of sustainability and sustainable development in Israel. He also serves as the Chief Sustainability Consultant to the Be'er-Sheva Municipality and regularly contributes opinion columns on environmental issues to various Israeli news outlets.

Chapter 1
The Origin of Matter

Abstract The "Big Bang," which began when an immense amount of energy collapsed and condensed into a singularity and then exploded, initiated the expansion of the Universe and defined the dimensions of Time and Space. The initial period following the "Big Bang" involved massive processes lasting millions of years, often described as the "evolution of matter." The first stage of matter formation, also known as "nuclear evolution," began with the compression of fundamental particles to form atomic components and the creation of light elements, such as hydrogen (H_2) and helium (He), which, through nuclear fusion, formed heavier elements like lithium (Li), beryllium (Be), and boron (B). These elements continued to serve as the bases for the formation of even heavier and more diverse elements, still used today in countless natural and artificial (human-made) processes.

Keywords "Big Bang" · Evolution of matter · Light elements · Hydrogen fuel · Lithium-ion battery

The creation of our Universe was initiated when energy became matter, as was described in the most famous equation in the world: $E = mc^2$, where the energy of matter at rest (E), equals mass (m) times the speed of light (c) squared (^{2}), as part of Einstein's principle of "mass–energy equivalence" in the theory of relativity [1]. This ongoing process of creation began ~13.7 billion years ago, when an enormous amount of energy that had been compressed and concentrated at a single point, a "singularity," exploded (the "Big Bang"), instigating the process of the Universe's expansion and cooling [2]. Universal expansion continued by means of a prolonged evolutionary process ("the evolution of matter", Fig. 1.1), which still shapes our environment today [3].

The first period following the Big Bang had huge processes that lasted millions of years and could be described as "atomic evolution," which included changes both in the structure and in the composition of the Universe. During the first period,

A. Wolfson, *Chemodiversity and the Ecological Crisis*, SpringerBriefs in
Molecular Science, https://doi.org/10.1007/978-3-032-07363-1_1

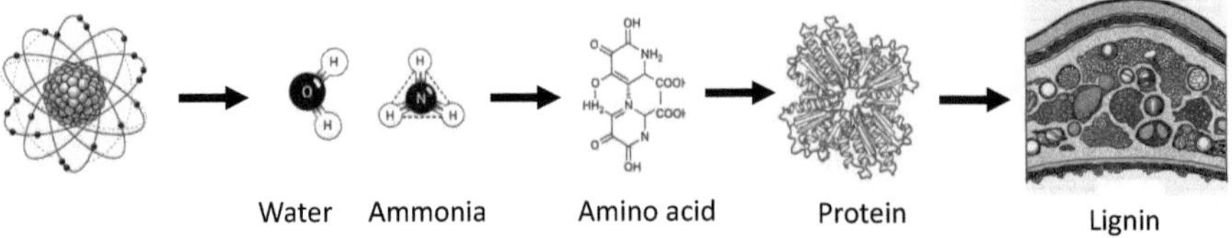

Fig. 1.1 Evolution of matter—from atoms to molecules to composite materials

which lasted ~370,000 years, the Universe was extremely hot and dense; however, as it expanded, it cooled, and its new conditions were suitable for the formation of various elementary and basic particles of matter that cannot be divided, termed "sub-atomic particles" (e.g., fermions, quarks, and leptons), and also "energy particles" or "force-carrying particles" termed bosons (e.g., photons and gluons)—which together constitute the building blocks of matter.

Meanwhile, and in relatively parallel amounts, particles of "antimatter" were being formed, having the same mass, electric charge and magnetic moment, but with opposite signs for their electric charge and magnetic moment. Thus, most of matter and antimatter quickly reacted together and were mutually annihilated, leaving only a small amount of matter in the newborn Universe. At this stage, the quarks aggregated forming "nucleons," which made up the nuclei of the various atoms—the protons carrying a positive electric charge (+1) and the neutrons (uncharged). The advent of negatively charged particles, the "electrons" (−1), took many more hundreds of thousands of years, at which time the electrons were captured in atomic orbits surrounding the existing nuclei, thus creating the first atoms of the different elements.

The Greek word "atom," meaning "cannot be cut or divided," was first coined in the fifth century B.C. by the Greek philosopher Democritus (370–460 B.C.), who hypothesized that all matter in the world is composed of one element that cannot be divided. Indeed, even today, knowing that different substances consist of more than one element, and that the 118 currently known chemical elements consist of smaller sub-atomic particles, they still cannot be divided by chemical methods and, therefore, they are still defined as "the basic units" of all matter. Ultimately, the distinct elements that were created via continuous evolutionary and energetic processes differ in the number of protons and neutrons in their atomic nuclei and in the number of free electrons orbiting them. In fact, these different amounts of sub-atomic particles determine the physicochemical properties of these differentiated elements.

The first stage in atomic evolution had included the formation of "light chemical elements," mainly hydrogen (H) and helium (He), and their various "isotopes" (each species of a certain chemical element shares the same number of protons and electrons but a different number of neutrons in its nucleus). However, some scientists also include three more elements and their isotopes in the group of light elements created minutes after the Big Bang: lithium (Li), beryllium (Be), and boron (B).

Hydrogen (H)

What do water (the liquid of life), food (ranging from carbohydrates and proteins to "lipids" (i.e., fats and oils)), nucleic acids (like DNA and RNA), fossil fuels (composed of hydrocarbons formed in the Earth's crust from decayed organic matter, such as *coal, oil, and natural gas*), and clothing (made of polyester) all have in common? On the one hand, they are all composed of molecules, essential for the existence and maintenance of human life and the fulfillment of our needs and desires; on the other hand, they all contain hydrogen on the molecular level.

Hydrogen, the simplest and lightest element in the Universe, is the first element in the Periodic Table. It is the most abundant chemical substance in the Universe, constituting ~75% of all "baryonic matter" (i.e., matter composed of subatomic particles, such as protons, neutrons, and electrons). In its atomic or ionic form as plasma, hydrogen is the main matter in intergalactic gas, stars, and interstellar clouds [4]. Contrarily, on Earth, the amount of free hydrogen (H_2), a colorless and odorless gaseous diatomic molecule with the lowest density of all the gases is very low, ~1400 ppm (parts per million) by weight (~0.14 wt%). In addition, its presence in the atmosphere is much lower than 1 ppm by volume, since hydrogen is so light that it cannot be contained by the planet's gravity; thus, when it forms, it escapes from the atmosphere into outer space. As such, hydrogen, which readily shares electron pairs (i.e., forms interatomic "covalent bonds") with most nonmetallic elements, is mainly found on Earth in water (H_2O) and its organic compounds—all with carbon (C) and hydrogen skeletons, accompanied by other elements, like oxygen (O), nitrogen (N), and others, and mainly as carbohydrates and hydrocarbons. Moreover, hydrogen, as a component of water and organic matter is also the third most massive element in the human body (comprising ~10 wt% of all the body mass).

Hydrogen has three stable isotopes: hydrogen with one proton in the nucleus, one electron orbiting the nucleus (marked as ^1_1H) and it is the most common among the isotopes (~99.99%); hydrogen containing one proton and one neutron in the nucleus and one orbiting electron is designated as ^2_1H or D, known as deuterium or "heavy hydrogen"; and hydrogen with one proton and two neutrons in the nucleus and one electron, denoted as ^3_1H or T, and called tritium. In addition, hydrogen also has some isotopes with a higher number of neutrons in the nucleus, but they are unstable. Deuterium is much less widespread than hydrogen, and it is believed that most of the deuterium was created at the outset of the Big Bang, while tritium, which is very rare in Nature, was mainly produced by the interaction of cosmic rays with atmospheric hydrogen.

The Hydrogen Cycle

Though molecular hydrogen concentration in the atmosphere, hydrosphere, and lithosphere is negligible, free hydrogen plays an active role in biological and ecological cycles. Gaseous hydrogen is used or produced in several biotic and abiotic processes, as part of the hydrogen cycle, where it is exchanged between living (biotic) and non-living (abiotic) surroundings via the production and decomposition of various molecules.

In the inanimate world, gaseous hydrogen is mainly produced by reactions between water and different minerals in the hydrosphere or lithosphere. Contrarily, the main "hydrogen sink" is caused by hydrogen being released into outer space. In the animate world, H_2 is produced or consumed by various microbial metabolisms using different enzymes. As such, "aerobic" (i.e., with the presence of oxygen) and "anaerobic" (i.e., in the absence of oxygen) microbial metabolisms consume H_2 by oxidizing it to reduce other compounds during respiration and fermentation, while producing H_2 from organic matter during the anaerobic bacterial metabolism of sugars, that yield organic acids. For instance, hydrogen can either be converted into its positive ions (H^+), which are actually protons, or generated from protons during reactions involving the enzymes—hydrogenases as catalysts and nitrogenases to reduce gaseous nitrogen to ammonia (NH_3); these two enzymes can also convert protons into hydrogen in the absence of nitrogen gas [3].

Gaseous hydrogen has also different man-made uses. Due to its low density, one of the first uses of hydrogen was for filling lighter-than-air airships—blimps and zeppelins. Yet, this use was soon abandoned in 1937, when the Hindenburg zeppelin caught fire, a catastrophe caused by a combustion reaction of hydrogen with oxygen. Nevertheless, hydrogen is still used today as lifting gas in weather balloons. Additionally, due to its relatively high thermal conductivity (compared to other gases), gaseous hydrogen is also used as coolant in certain type of generators. Moreover, currently, hydrogen is mostly used as an "industrial feedstock" (i.e., raw material) to produce various reduction reactions—in which molecules, atoms, or ions gain electrons or decrease their "oxidation states" (representing the number of electrons an atom can gain, lose, or share when chemically bonding with an atom of another element).

Since hydrogen is a very reactive element that can both release and receive electrons, it reacts with most elements. When a hydrogen atom reacts with a more electronegative atom than itself, it attracts a shared pair of electrons, such as oxygen or chlorine (Cl), forming covalent bonds—and it has a positive charge. Contrarily, when in the presence of less electronegative elements than itself, like metals, such as sodium (Na), it forms a hydride (i.e., a hydrogen anion, H^-). Furthermore, as in natural reactions, hydrogen can also be used to form acids or bases. The reaction of hydrogen with nitrogen to produce ammonia has frequent uses in agricultural fertilizers and explosives. Yet, the greatest utilization of gaseous hydrogen is as a cooling agent (employing over 50% of all the hydrogen production worldwide). It is also used to reduce many unsaturated organic compounds with double or triple carbon-carbon bonds that are highly prone to form atomic bonds with other atoms. This can be accomplished either by using metal hydrides, or by performing hydrogenation with molecular hydrogen, or by "transfer hydrogenation" (in which hydrogen is transferred from one organic molecule to another); for example, the partial hydrogenation of oil or fat to make margarine, i.e., the reduction of unsaturated fatty acids to their corresponding saturated forms. In addition, hydrogen is also used in the production of synthetic fuel by means of the Fischer-Tropsch reaction, in which synthesis gas, or "syngas," consisting of a mix of hydrogen and carbon monoxide (CO), reacts to form organic compounds. Finally, hydrogen also has various other industrial applications—as an inert and/or protective gas in flat glass production and as a protective and/or "carrier gas" in the electronics industry.

Hydrogen also plays active role in hydrogen or thermonuclear bombs, which are even more powerful than atomic bombs. An atomic bomb explodes when an atom's nucleus (e.g., plutonium (Pu) or uranium (U)) is split in a fission reaction, releasing neutrons (neutral particles), that hit the other nuclei of nearby atoms, splitting them too—thus yielding an explosive, high-energy chain reaction. In hydrogen bomb, on the contrary, a combination of fission and fusion occurs, in which the heat generated by the fission of plutonium or uranium causes deuterium-tritium fusion (D + T), that then go on to resale neutrons, further boosting the explosive chain reaction of plutonium or uranium.

Hydrogen Fuel

Electrochemical cells, consisting of metallic anodes and cathodes, plus "electrolytes" containing salts or acids, able to conduct electric currents using trace amounts of gaseous hydrogen in the air, combined with oxygen and burnt, to release both energy and water, were already known by 1842. A more efficient process for the conversion of hydrogen and oxygen into water and the continuous-mode generation of electricity, i.e., the "Grove voltaic cell," was first proposed by William Robert Grove (1811–1896), an apparatus that later became known as a "fuel cell." Yet, the use of these newly-developed hydrogen fuel cells required the use of pure gaseous hydrogen, which was not just expensive, but also very volatile and potentially explosive; not to mention the internal reaction within those batteries that also use dangerous nitric acid (HNO_3) as an electrolyte and discharge poisonous nitrogen dioxide (NO_2).

Meanwhile, in the late nineteenth and twentieth centuries, liquid fossil fuels, mainly crude oil, being used more and more as they became much cheaper, joining the extensive use of coal in production of electricity, making the fuel cells and hydrogen batteries passé. Yet, over a hundred years later, with the ongoing efforts to phase out fossil fuels, due to their release of greenhouse gases (GHGs) and air pollutants during their extraction, conduction, storage, and combustion, and to deal with climate changes—hydrogen is once again being recognized today as an important future energy.

Now, in the twenty-first century, hydrogen, in its liquid form, which reacts strongly with oxygen, is being used as a fuel primarily in rockets. However, this simple and very spontaneous energetic reaction is now of great interest to the energy sector that is using fuel cells to convert chemical potential energy into practical electric energy. Fuel cells provide many advantages, as they are more efficient than combustion engines, and are much more environmentally friendly. Currently, hydrogen is a much more renewable and sustainable fuel than oil, which requires extraction and refining. Moreover, the combustion of hydrogen does not pollute the environment, leaving only water—thus, avoiding the release of carbon dioxide pollution (the GHG that fuels the climate crisis) and other air pollutants that cause severe health risks.

Despite its promising features, using hydrogen as fuel also has some limitations, since it requires a specific type of secure packing and storage, which is not trivial, and its handling requires serious safety precautions. In addition, as a gas, hydrogen has low density, so it must be compressed, even liquefied, for storage, using a great deal of energy. Furthermore, since hydrogen is both very explosive and highly flammable when mixed with ordinary air even in small amounts, all the processes that employ hydrogen must work with unique and relatively expensive systems, under very strict safety regulations. Thus, extensive daily use of hydrogen will eventually lead to different systems that generate hydrogen on location from metal hydrides, organic molecules, and more.

Besides the storage and handling safety aspects, the main issue with the use of hydrogen as a fuel is its production—considering both the source of the hydrogen and the source of the energy to be used in its production process. To-date, several types of hydrogen have been distinguished on the "hydrogen color palette," classified by its source substance and method of production: "black" and "brown hydrogen" produced from black (bituminous) or brown (lignite) coal, respectively, during a process with high emissions of carbon dioxide and methane (CH_4)—both GHGs. "Gray hydrogen" is produced from natural gas by reforming the methane with steam, yielding fewer carbon dioxide emissions, but more methane emissions with much higher global warming potential (GWP) than carbon dioxide. "Blue hydrogen" is produced from fossil fuels, but its production combines capture and storage of GHGs (carbon capture and storage (CCS)). "Turquoise hydrogen" is produced from methane in a pyrolysis process at high temperature, and pure carbon is produced alongside the hydrogen. "Green hydrogen" is separated from water by electrolysis, a process that splits water to gaseous hydrogen and gaseous oxygen by electric current, therefore the only by-product in the process is oxygen, when the production's energy is from renewable energy. When solar energy is used, it is often referred to as "yellow hydrogen". Moreover, the production of hydrogen from water using renewable energy makes it possible to use hydrogen also to store excess renewable energy, which its production is often irregular and depends on time and place. "Pink," "purple," and "red hydrogens" are derived from water using nuclear energy. Natural hydrogen produced from underground deposits or from the air is termed "white hydrogen."

Finally, since there is also shortage of "freshwater," the production of hydrogen fuel from seawater has the notable advantage. Another environmental concern is the increase of molecular hydrogen emissions into the atmosphere, which have a negative effect on the Earth's climate. In fact, hydrogen can be called an indirect GHG, because it induces perturbations in atmospheric methane and ozone (two potent GHGs), by reacting with hydroxyl radicals (•OH), while producing stratospheric water vapor (another GHG).

Helium (He)

Helium is the second element in the Periodic Table and the second element in abundance in the Universe after hydrogen, composing ~25% of all baryonic matter, being a major component of stars, where it was created by the fusion of two hydrogen atoms [5]. It is found in the air, the sixth molecule on a "dry basis" but being very light, it mostly diffuses into outer space and, thus, its concentration in the lower atmosphere is only ~5 ppm. Helium is a light, colorless, odorless, "monatomic gas," with a complete outer electron energy level surrounding its atomic nucleus, containing a specific number of electrons. As such, helium is also the first of a group of "noble gases," all having full outer shells and very low reactivity. These gases are also referred to as "inert gases," due to their very low reactivity with other substances.

Helium has no biological activity or function, yet it has many medical applications, owing to its distinct physical and chemical characteristics, namely: its low density, low solubility, and high thermal conductivity [6]. For example, a harmless, breathable mixture of two gases, named Heliox, is a mixture of helium and oxygen, in which the helium replaces nitrogen; since helium is much lighter than nitrogen, the use of Heliox improve the flow of oxygen in patients suffering from various respiratory problems, like asthma or bronchitis. This artificial atmosphere is also used by deep-sea divers ("diluent gas") and others working under pressurized conditions. Helium has also been used in the treatment of the cardiovascular system and may aid in the protection of myocardial tissue against ischemia (when the blood supply to the tissues is restricted). There is also a neuroprotective treatment for brain injury using helium under high pressure and some new surgical applications. Finally, with a boiling point of minus 268.9 °C, liquid helium is the coldest element on Earth and its most common use in medicine is not in the body, but rather as the coolant in "magnetic resonance imaging" (MRI), which operates utilizing a superconducting magnet. Similarly, helium is used in spectroscopy to cool *"nuclear magnetic resonance"* (NMR) spectrometers, which a key technique for detecting molecular structures. Furthermore, when using a "scanning helium ion microscope" (SHIM) for imaging, a helium ion beam is used to get sub-nanometer resolution.

Helium has many other uses in different other fields, as well. One of its earliest uses was (and still is) for the floatation of airships, filling balloons, and inflatable air bags in cars. Currently, it has a variety of applications in semiconductor manufacturing, from coolant to diluent gas in etching processes. It is also employed in the manufacture of fiber-optic cables, used for Internet and television, in which a helium atmosphere is supplied to avoid the formation of trapped air-bubbles inside the cables. In addition, helium is used as: a filler in hard drives, instead of air (resulting in much higher storage capacity with reduced power expenditure); in helium-neon (HeNe, another noble gas); and in gas lasers. Moreover, helium is massively applied for detecting leakage, thanks to its quality of rapid diffusion. Finally, super-liquid helium is also the cooling medium of the giant magnets that help to shoot particles inside the Large Hadron Collider (LHC), the world's largest and most powerful particle accelerator, where an international team of researchers is trying to uncover the composition and mechanisms of the Universe.

On Earth, helium is stored underground, where it was created by the decay of radioactive elements in rocks. This process includes the emission of "alpha particles" consisting of two protons and two neutrons, like helium's nucleus, which later gains two electrons. Thus, most of the helium in the world is found in natural gas fields, where it is mixed with different hydrocarbons and other substances; harvesting helium is mainly done by means of a "fractional distillation" separation process. However, in these raw mixtures, there is only a small amount of helium; even in the richest fields, it makes up less than 10 vol% of the total volume of the gases at that location. Moreover, it is impossible to produce high amounts of helium synthetically.

Helium, itself, is neither an air pollutant nor a GHG, yet its exploitation from gas fields causes high emissions of methane, and the liquification of helium for storage and transportation requires a high amount of energy, usually produced by the combustion of fossil fuels that do generate GHGs and air pollutants. In addition, though helium has no known biological or ecological function, it is an essential and irreplaceable commodity for the manufacturing of many man-made products and numerous applications. However, due to its massive exploitation over the years, we are running out of helium. In fact, the relatively large amount of helium that has been extracted during natural gas production and its relatively low price have made its recovery and recycling seem redundant. Nonetheless, some experts warn that the world could run out of helium within 25–30 years; thus, they suggest formulating a new helium exploration and storage policy with an updated price tag now.

Lithium (Li)

Lithium is the third element in the Periodic Table, and the first one in the group of metals—mostly solids, characterized by a lustrous appearance and high electrical and thermal conductivity, due to "metallic bonds," that is, electrostatic attraction between metal "cations" (positively charged ions) and delocalized electrons [5]. It is the first in the column of alkali metals, all shiny, soft, and highly reactive. Under normal conditions, lithium is the metal with the lowest density and is the lightest solid. Though lithium has several known isotopes, only two of them—^7_3Li with 4 neutrons in the nuclei (~90%) and ^6_3Li with 3 neutrons in the nuclei (~10%)—are stable.

Lithium was created shortly after the Big Bang, along with hydrogen and helium, yet it can be destroyed at relatively low stellar temperatures when colliding with a proton to yield two atoms of helium. Thus, it is ranked as the 25th among the 32 first chemical elements in the Solar System [7]. Due to its reactivity, lithium does not exist on Earth in its metallic form, though it is found in Earth's crust within different compounds, like *salts* and *minerals*, which appear in rocks in the lithosphere and in soil. Furthermore, some lithium has been found in the form of a soluble ion in the hydrosphere.

Lithium has many uses in different fields. Though its biological functions and necessity in living organisms have not been fully researched, it shows biological activity in biotic environments, including the human body, where it is considered a "trace mineral," essential for regulatory functions. Additionally, though lithium is not known to be essential for plants, small amounts of lithium in plants seem to have some positive effects; however, high concentrations were found to be toxic [8]. Similarly, it was also discovered that lithium may have some beneficial and some detrimental effects on animals and human beings. It can affect healthy brain function and performance, since its most important roles in the body is the regulation of serotonin—a naturally occurring neurotransmitter that carries signals between nerve cells and, among other things, also positively affects people's moods. As such, it is frequently used in psychiatric medication, when treating, for example, manic-depressive disorders and schizophrenia. Lithium was also found to improve cognitive function, focus, and attention, and to improve the cardiovascular and immune systems. Nevertheless, lithium can also produce negative effects and toxicity due to the replacement of sodium and potassium cations (K^+) to inhibit certain enzymes and biological cycles.

Lithium has many nonbiological applications. It was first used in 1938 as grease for mechanical lubrication, for instance in aircraft engines, since it is less corrosive than other alkaline-based soaps. It is also used in nuclear weapons, like hydrogen bombs, because neutron irradiation of both lithium isotopes yields tritium. Lithium metal is also added to magnesium (Mg) and aluminum (Al) alloys to improve their strength while making them lighter. High amounts of lithium are often used in the production of ceramics and as glass additives, as well as functioning as a desiccant in industry. Finally, lithium salts are used in a variety of applications: air purification, organic chemistry, the polymer industry, as well as having applications in the field of optics. The most common use of lithium nowadays is in lithium-ion batteries, frequently used in personal electronic devices: mobile phones, laptops, etc., and in most electric cars, thanks to its high energy density. Thus, the industrial demand for lithium is expected to increase exponentially over the coming decades. Yet it is estimated that the current global reserve of lithium is high and will be enough for the foreseeable future.

Most lithium is extracted from lithium rich brine, by means of the solar evaporation of water, although lithium is also extracted from minerals by a process of electrolysis. The disadvantage of lithium extraction is that it inevitably harms the soil and causes air pollution, potentially becoming an environmental contaminant with short- and long-term effects on the animate world, including Humanity and the natural ecosystems.

Furthermore, the relocation of lithium from its natural habitats into urban industrial "effluents" from lithium-ion battery factories or devices containing lithium wastes, thrown into municipal dumps, may cause water and soil contamination and other negative environmental impacts.

Boron (B)

Boron is a metalloid with mixed characteristics of metals and nonmetals. It is a sparse element (i.e., neither abundant nor rare), having only 10 ppm by weight. It is not usually found in Nature in a pure state, rather it occurs as a component in various minerals [5]. One such mineral is borax ($Na_2B_4O_7 \cdot 10H_2O$), also known as sodium borate or sodium tetraborate, a white boron mineral used many years ago, for different purposes, ranging from cleaning to pottery glazes and in the process of mummification.

Boron is an essential nutrient for plants, necessary for the building and maintenance of their cell walls [9]. However, like many nutrients, a high level of boron can harm plants, causing decreased growth and productivity. Since it is found in the food chain of plants, boron is also consumed as food by animals, including human beings, in small amounts thought to have no toxic effects. In addition, though there is some evidence that boron is an essential element for both animals and humans, serving bone development, brain metabolism, and the immune systems, scientific knowledge of these functions is minimal. Although low-levels of boron-based minerals are non-toxic in humans and animals, high boron uptake, from food and water or due to a polluted environment, can cause negative effects to the stomach, liver, kidneys, and brain, and may result in death [10].

Boron, boronic acid, and minerals of boron are used in some manufactured goods, such as: fibers, glass, detergents, ceramics, and fertilizers. Boron is also a "doping agent" (which alters certain physical properties) in semiconductors, of silicon or germanium (Ge). Moreover, it is used to synthesize various molecules in organic chemistry, in catalysts, and in solvents, etc. Boron and borax are also used in household dishwashers, and as industrial detergents, cleaning products, and bleaching agents, and so they are found in both residential and industrial wastewater, which can contaminate nearby bodies of water.

References

1. Einstein A (1919) Time, space, and gravitation. London Times, London
2. Eden AH, Moor JH, Soraker JH, Steinhart E (eds) (2013) Singularity hypotheses: a scientific and philosophical assessment. Springer, Berlin
3. Chaisson EJ, Chaisson E (2001) Cosmic evolution: the rise of complexity in nature. Harvard University Press, London
4. Rigden JS (2003) Hydrogen: the essential element. Harvard University Press, London
5. Stwertka A (1996) A guide to the elements. Oxford University Press, London
6. Berganza CJ, Zhang JH (2013) Med Gas Res 3:18
7. Lodders K (2003) Astrophys J 591:1220
8. Shahzad B, Mughal MN, Tanveer M, Gupta D, Abbas G (2017) Environ Sci Pollut Res 24:103–115
9. Kabu M, Akosman MS (2013) Biological effects of boron. In: Reviews of environmental contamination and toxicology. Springer, New York, pp 57–75
10. Bolan S, Wijesekara H, Amarasiri D, Zhang T, Ragályi P, Brdar-Jokanović M et al (2023) Sci Total Environ 894:164744

Chapter 2
The Variety of Elements

Abstract The variety of elements refers to the wide range of chemical elements that exist, each defined by a unique number of protons in its atomic nucleus. These elements—formed during the early stages of the Universe and by nuclear processes within stars—differ in their physical and chemical properties, due to variations in their atomic structures. This diversity allows for the creation of countless compounds and materials essential to both natural processes and artificial (human-made) systems. In the presence of high energy, atoms bonded through a process known as "chemical evolution" or "prebiotic evolution," forming simple organic and inorganic molecules, such as methane (CH_4) and ammonia (NH_3), laying the foundations for the emergence of life. In addition, Humanity has created synthetic substances from these elements—many of which do not decompose naturally and, thus, contribute to pollution, disrupting the function of natural ecosystems.

Keywords Chemical evolution · Organic compounds · Inorganic compounds · Pollution

At some time between 100 and 250 million years after the Big Bang, the first stars appeared. They mostly consisted of dark matter—matter that is invisible to the naked eye and does not absorb, emit, or incite light, hence its name. Many believe it consists of elemental particles or a type of matter still unknown to us, that has gravity, but that causes different sorts of interactions with hydrogen and helium than those with which we are currently familiar. Under the high temperatures and pressures in the cores of the stars, these elements became ionized (i.e., lost an electron), and underwent sequences of nuclear fusion reactions, yielding heavier elements—such as lithium, beryllium, boron, carbon, nitrogen, oxygen, phosphorus (P), and sulfur (S)—that together with hydrogen form the most common elements in living organisms and in the biosphere.

Alongside the most common elements in living organisms (that dominate the biotic environment), there are also different metallic and nonmetallic elements that dominate abiotic surroundings, which were formed by the nuclear fusion within the

A. Wolfson, *Chemodiversity and the Ecological Crisis*, SpringerBriefs in Molecular Science, https://doi.org/10.1007/978-3-032-07363-1_2

stars' cores, for instance: sodium, potassium (K), and chlorine, commonly dissolved as ions in the hydrosphere. Aluminum, silicon, and iron (Fe) form the basic compounds that compose the Earth's "geosphere" and "lithosphere."

While the elements up to iron (atomic number 26) are created by fusion in the cores of stars, heavier elements need harsher and more extreme conditions to be created during a process called "rapid neutron capture." In this process, within a few seconds, atomic nuclei are bombarded by neutrons and destabilized before they decay naturally; this yields heavier elements, ranging from iodine (I, at. no. 53) to uranium (U, at. no. 92). This process can occur when a star's life ends as it reaches "critical mass"—when all its nuclear fuel is spent (i.e., no elements are left to that can fuse into heavier elements, so there is no internal force or pressure remaining to counterbalance its self-gravity)—causing an implosion termed "core-collapse," or when a massive star explodes, termed a "supernova," usually due to "thermonuclear runaway" (i.e., the emission of a huge neutron flux, in which heavy elements are built up by successive neutron absorptions, termed "nucleosynthesis"). After the explosion, the dust and gases of the dead star spreads in the interstellar medium (the region between the stars) that contains mainly clouds of gases, composed of hydrogen, helium, and minute solid particles. These huge clouds of gas and dust are called "nebulae," and when they reach a certain density, their gravity pulls the cosmic dust and gases inward, and they slowly collapse into a number of very hot cores. When these cores become hot enough for nuclear fusion to occur, they become self-powered newborn stars that power themselves.

The Abundance of Chemical Elements

The types and compositions of the elements, atoms, and molecules are a couple of the footprints that distinguish between different biotic and abiotic environments. The most abundant elements in our Milky Way Galaxy, our atmosphere, hydrosphere, lithosphere, and biosphere are listed in Table 2.1.

Table 2.1 The most abundant elements in abiotic and biotic environments

	Milky Way [1]	Atmosphere [2][a]	Hydrosphere [3]	Lithosphere [4]	Biosphere
1	Hydrogen (73.9)	Nitrogen (78.4)	Oxygen (88.8)	Iron (32.1)	Oxygen (65)
2	Helium (24.0)	Oxygen (21.01)	Hydrogen (11.1)	Oxygen (30.1)	Carbon (18)
3	Oxygen (1.0)	Aragon (0.46)	Chlorine (<1)	Silicon (15.1)	Hydrogen (10)
4	Carbon (0.5)	Carbon (<1)	Sodium (<1)	Magnesium (13.9)	Nitrogen (3)
5	Neon (0.1)	Neon (~0)	Calcium (<1)	Sulfur (2.9)	–
Other	0.5	~0	~0	5.9	7

[a] Excluding water vapor

Carbon (C)

Carbon is the main basic element among the organic compounds that compose all organisms and, hence, the biotic environment. It is the fourth most abundant element in the Universe and is the second most abundant element in the human body by mass (~19 wt%), also found in high amounts in the Earth's crust.

Carbon has been known since antiquity in its form as charcoal (i.e., light-weight black carbon produced by heating wood) and was already used ~6000 years ago as fuel and in the production of different metals, such as in the process of manufacturing bronze—an alloy of copper with tin (Sn) and other metals, like zinc (Zn) or nickel (Ni), or nonmetallic phosphorus. Later, charcoal was also found to have various medicinal uses and applications as an adsorbent. Additionally, it was also recognized that the particles in exhaust gases emitted by combustion processes, termed "soot," are also made of carbon. Furthermore, graphite and diamonds, which are carbon "allotropes" (i.e., chemical elements that exist in different forms), have also been identified and used for many years, although no one thought these allotropes all consisted of the same element. Yet, graphite and diamonds, are not the only allotropes of carbon, and include fullerenes, graphene, carbon fibers, and carbon nanotubes—all currently serving in many special applications. These new materials have novel properties and many potential applications in nanotechnology, electronics, optics, and other fields of materials science. Finally, their full combustion with oxygen yields gaseous carbon dioxide (CO_2), while their partial combustion yields carbon monoxide (CO)—both of which interact with water, producing carbonic acid (H_2CO_3). Like hydrogen, carbon dioxide is also a key source of all organic compounds.

The natural production and consumption of carbon-based matter or organic compounds is illustrated by the fast and slow carbon cycles. The fast or "biological carbon cycle" is initiated by photosynthesis, where carbon dioxide and water react in the presence of sunlight to yield sugar, as building blocks and energy sources in the biosphere. All the organic compounds that are produced in photosynthetic organisms pass through the food chains from plants and algae to herbivorous and algivorous, i.e., plants and algae consumers, correspondingly, and from them to carnivorous, i.e., meat consumer. At the same time, alongside respiration processes in animals and plants, carbon dioxide is also released by the burning of organic matter or in the decay of plants and animals in an oxygen-rich environment, i.e., an aerobic environment, as well as in natural weathering processes of rocks. Today, however, the bulk of the carbon dioxide released into the atmosphere is accompanied by water, as byproducts of the burning of fossil fuels, such as coal, petroleum, and natural gas, by humans.

The movement of carbon between the biosphere, hydrosphere, lithosphere, and atmosphere is described by the "slow carbon cycle" or "geochemical carbon cycle." Unlike the fast cycle, which occurs on orders of magnitude of tens to hundreds of years and moves 10^{16}–10^{17} g of carbon (100,000–10,000 million tons) per year, the processes during the slow cycle continue over tens and hundreds of millions of years, and the amount of carbon movement is smaller: $10^{13}-10^{14}$ g of carbon (10–100 million tons) per year [5]. The movement of carbon in the slow carbon cycle begins with carbon dioxide from the atmosphere that dissolves in the rainwater and forms a weak carbonic acid, which falls with the rain into the bodies of water and on the ground. The carbonic acid that reaches the ground causes the chemical weathering of various rocks, releasing minerals and metals as cations of sodium, potassium, and magnesium, as well as calcium (Ca) ions, that are washed into the streams, rivers, seas, and oceans. The carbonic acid that reaches the oceans turns into bicarbonate (HCO_3^-) and carbonate (CO_3^{2-}) ions, that, in turn, react with calcium ions to form calcium carbonate ($CaCO_3$) that is the main mineral in various sedimentary rocks and the main component of the skeletons of various marine organisms. Finally, the completion of the slow carbon cycle occurs when carbon dioxide is deposited into the atmosphere via the emission of gases from volcanoes and by the processes of combustion and decay of organic matter. In order to maintain a carbon balance, the total amount of carbon in each of the reservoirs should be more or less constant, and the flow of carbon moving to and from this or that reservoir should be more or less the same.

Human activity has tremendous effects on both carbon cycles, which in turn effect Humanity. The combustion of fossil fuels and the growth of food, and especially sheep and cattle, lead to the release of greenhouse gases (GHGs) (mainly carbon dioxide and methane). The reassignment of land use for urban purposes and "deforestation" (the clearing of forests), along with the production of concrete from limestone, are also responsible for changing the balance between the different components in the carbon cycle. Many organic compounds, ranging from volatile compounds, to oils, and from fats to polymers (among them a variety of synthetic compounds) contaminate the air, the land, and the water sources. These alterations to the carbon cycles, alongside the emissions of many carbon-based materials into the environment, all produce serious impacts on the climate, biodiversity, and the pollution crises, as well as the whole of Nature, and, ultimately, on the survival of Humankind.

Nitrogen (N)

Nitrogen, the seventh element in the Periodic Table, is also the seventh most abundant element in the Universe, the fourth most abundant element in the human body (3 wt%), and the most widespread element in Earth's atmosphere or air (78 vol%), where it appears in the form of a diatomic molecule (N_2).

Nitrogen has many biological uses. It is an essential element in the amino acids that serve as building blocks in the protein-production process. Proteins are key components in all organisms and are responsible for many regulatory processes in plants and animals. Nitrogen is also a central component of chlorophyll, the most important pigment used for photosynthesis, as well as in adenosine triphosphate (ATP), an important "energy molecule" found in all life forms. In Nature, the nitrogen cycle is a major biogeochemical cycle that supports the existence of the biotic environment. It begins with molecular nitrogen. However, since gaseous nitrogen cannot be absorbed directly into biosystems, it is transformed by bacterial and algal systems into nitrate ions (NO_3^-), either directly or by the production of ammonia (NH_3), which then undergoes nitrification processes, becoming ammonium ion (NH_4^+), nitrite ion)NO_2^-), and nitrate ion. Nitrate is the main element that is assimilated by plants for the creation of amino acids and plant proteins, which are the main source of proteins in animals. The nitrogen assimilated by the proteins is finally broken down in animals' bodies (mainly by the liver) into urea (CH_4N_2O) or uric acid ($C_5H_4N_4O_3$), and then further broken down by water into ammonia or the ammonium ion. Moreover, when microorganisms, animals, and plants die, their proteins break down as well, releasing ammonia or nitrogen gas back into the atmosphere. In addition, nitrogen has several significant industrial uses, ranging from the provision of an inert atmosphere in the food and electronics industries, the annealing of stainless steel by heating (by making it more malleable), to the production of fertilizers, nitric acid, nylon, dyes, and explosives, beginning with the production of ammonia by means of nitrogen's reaction with hydrogen.

Though nitrogen is an inert gas, it may somehow affect human health and the environment. Excess nitrogen in the air can impair our ability to breathe, limits visibility, and alters plant growth. Yet, the main hazardous effects of nitrogen are due to emissions from various nitrogen-based compounds, such as ammonia, which, if released into the air, can cause irritation, burning of the skin, mouth, throat, lungs, and eyes, even serious damage to lungs and death. Various nitrogen-based fertilizers may emit ammonia into the soil and water sources, thus causing a nutrient imbalance impacting entire ecosystems. Furthermore, nitrogen oxides (NOx), air pollutants emitted by the burning of fossil fuels, increase the risk of causing respiratory problems, resulting from inflammation of the lungs' lining and/or reduced immunity to pulmonary infections, characterized by wheezing, coughing, colds, flu, bronchitis, and asthma attacks. Additionally, the nitrous oxide (N_2O), also known as "laughing gas," used in hospitals and dental clinics for its ability to dull pain, is mostly released by natural sources (e.g., soils under natural vegetation and the oceans), while its human sources come from agriculture, mainly from the decomposition of excess fertilizers, fossil fuel combustion, and industrial processes. This nonirritant and nontoxic molecule is, nevertheless, a very aggressive GHG, with very high global warming potential (GWP).

The dispersion of nitrogen oxides in the atmosphere leads to their absorption in cloud water droplets forming "acid rain." Acid rain precipitation, or any other precipitation form, is quite damaging to plants and animals, their habitats, and ecosystems. Acid rain can directly damage human health, as well as dissolving of certain rocks and doing damage to buildings and infrastructures by corrosion. Acid rain is harmful to forests, and, when it seeps into soil, it dissolves the essential nutrients (like magnesium and calcium) that maintain the health of the tree. Acid rain negatively affects lakes, streams, wetlands, and the biodiversity of their surrounding habitats.

Oxygen (O)

What do we need to stay alive? Oxygen! A few minutes without oxygen can cause permanent brain damage and a few more minutes lead to death. Most living things (bacteria, animals, and plants) consume oxygen to survive, either by breathing atmospheric dioxygen (O_2)—a simple and very widespread gaseous molecule—or by separating dissolved oxygen from water.

Oxygen, the eighth element in the Periodic Table, is a nonmetal that makes up ~21 vol% of the atmosphere, ~49 wt% of the Earth's crust, and ~ 67 wt% of the human body. It circulates within biotic and abiotic environments, while reacting with different elements and compounds, sometimes transforming within one of the main biogeochemical cycles—the oxygen cycle.

The oxygen cycle starts with atmospheric molecular oxygen adsorbing into the flora and fauna in the biosphere by means of natural respiration processes that create energy by means of the decomposition of organic matter, by releasing carbon dioxide. In a similar process, molecular oxygen serves as the main oxidizing agent during the neutral combustion of organic matter, in which carbon dioxide and water are also emitted along with gaseous sulfur oxides (SO_2, SO_3), nitrogen oxides (NO, NO_2), and carbon monoxide (CO). Oxygen is also the primary components in various solid oxides (very common in the solid matter of the Earth), such as silica, alumina, and lime, created by reactions between a metal and oxygen or water. Then, molecular oxygen is emitted, returning to the atmosphere during photosynthesis. Oxygen also serves as a key element in many man-made processes, from combustion of fuels to create energy to chemical and metallurgical processes. Finally, completing the oxygen cycle loop, oxygen also naturally appears in the form of ozone gas (O_3), especially in the very important atmospheric layer of "good ozone," protecting the Earth's surface from dangerous ultraviolet (UV) radiation.

Though the decrease in the concentration of oxygen in air is not crucial and does not pose a threat, the change in the oxygen concentration in oceans, seas, and other bodies of water, is of great concern. For many years, the concentration of oxygen in the oceans and seas has been constantly diminishing, though it cannot be seen by the naked eye. There are several reasons for this phenomenon, all due to human deeds, especially global warming, as the warmer the water, the less oxygen it dissolves. Various recent studies show that, over the last 60 years, the total oxygen concentration in the oceans has diminished by ~2% on average. This may not seem like a dramatic change, but this decrease is not uniform everywhere, such that there are certain areas in the oceans that have already lost 10–40% of their oxygen.

Phosphorous (P)

Phosphorus is a nonmetallic element, the fifteenth element in the Periodic Table. It has several allotropic forms: "white phosphorus," which has a wax-like structure of four covalently-bonded phosphorus molecules (P_4), produced industrially by heating phosphate (PO_4^{3-}) rock in the presence of carbon and silica in the oven; "red phosphorus," produced by heating white phosphorus in an oxygen-free atmosphere, to form an amorphous form; and "black phosphorus," produced by heating under pressure and creates a structure of plates like in graphite. White phosphorus is a toxic solid, and in contact with the skin can cause severe burns. It glows in the dark and it is spontaneously flammable when exposed to air. As such, white phosphorus is used as an igniter for various bombs, and in phosphorous bombs, an incendiary weapon that in addition to their destructive explosive power can spread fire and cause injuries that are more serious and harder to treat than injuries from conventional bombs. Red phosphorus in contrast is a nontoxic solid used, for example, on the sides of matchboxes to light them. Phosphorus also has uses in the field of semiconductor materials, in steel production, in cleaning agents and in insecticides, and in the food industry as an additive that preserves the texture of cheese and frozen meat, and in different beverages. Compounds containing phosphorus are used also in the production of porcelain and various alloys. But the biggest use of phosphorus compounds is undoubtedly in the fertilizer industry, as phosphorus is an essential element in all components of the biotic environment. Phosphorylation is an essential component of DNA and RNA, and the ATP that transfers energy in cells. Additionally, phosphorus is also found in many other biologically important molecules, and it also helps in maintaining the acidity level in the body.

The average daily phosphorus consumption of a person is ~1 g of phosphate per day, of which ~750 mg are assimilated in the body, mainly in the bones and teeth. The primary source of phosphorus for all animals, including man, is food, especially meat, eggs, nuts, and seeds, including grains.

Phosphorus appears in Nature mainly in sedimentary rocks that are rich in phosphate. Although the phosphate itself can be used as a fertilizer, most fertilizers today are produced using a synthetic method that mixes three minerals, which are essential to plants: potassium, nitrogen, and phosphorus. As such, the phosphorus extracted from the phosphate is used to produce phosphoric acid (H_3PO_4), which is the main source of phosphorus in fertilizers.

Phosphorus is a limited resource, and its global reserves are concentrated in a small number of countries. Over the years, various studies warned that once the existing amount of phosphorus is enough to satisfy all the human demands, this natural resource will already be significantly depleted (i.e., there will be a shortage of natural phosphorus). Moreover, since it is an essential resource for sustaining life, its shortage and increased price will affect food security in the entire world. Therefore, it is important to conserve the phosphate reserves, locally and globally, if we wish to satisfy the needs of future generations, as well as stopping the rising cost of phosphate.

We find ourselves in a quandary, because the use of phosphorus in fertilizers is, on one hand, indeed necessary for the Humanity's food security, while, on the other hand, the mining of phosphate, the production of fertilizers, and their hazardous residues pollute our equally vital air, soil, and water.

One of the main problems with phosphate and other similar, essential minerals, is that the growth of the plant that absorbs the minerals from the soil happens in one place, while the transportation of the goods and the disposal of the waste products, containing some of those minerals, occur elsewhere. Note that phosphorus, unlike nitrogen, does not evaporate into the air or dissolve in water in large quantities, and therefore it does not move from place to place. As such, "the phosphorus cycle" (another "biogeochemical cycle"), in which phosphorus transfers from rocks to plants, alga, and animals, in the sea and on the land, is disjointed—it does not close properly, as in the cases of the carbon, nitrogen, and oxygen cycles.

In recent years, phosphorous and phosphate have found new and promising uses in batteries for energy storage, whether for powering vehicles or in renewable electricity production facilities, such as solar fields and wind turbines. In these batteries, known as "lithium-iron-phosphate (LFP-LiFePO$_4$) batteries," the addition of iron and phosphate to the lithium-ion cathode yields improved performance and additional benefits, and the quantity of these batteries is expected to swell in the future. However, the use of phosphorus in batteries raises an ethical issue regarding the diversion of phosphate from agriculture to the battery market.

Sulfur (S)

Sulfur, the sixteenth element in the Periodic Table, is a nonmetal that has been known since ancient times and is mentioned in the Hebrew Bible. It is a soft, pale yellow, odorless, brittle solid, insoluble in water that makes up ~3% of the Earth's mass. It is found in many allotropes, and also has different polymorphs, the most common form of which is the "orthorhombic polymorph" of S$_8$, with a ring or "crown" structure.

Sulfur has a unique biological role and is essential to all living things. It is one of four amino acids: methionine, cysteine, homocysteine, and taurine, though only the first two are incorporated into proteins, in which sulfur bridges between two amino acids, forming and stabilizing the 3D design of proteins. Sulfur passes through a biogeochemical cycle, in the first phase of which sulfate (SO$_4^{2-}$) from the soil or in water reserves is consumed by organisms, upon the death of which that sulfur is released back into the atmosphere as hydrogen sulfide gas (H$_2$S). In the second phase of the sulfur cycle, the single-celled organisms, living in seabed sediments, use the sulfur dissolved in the seawater to digest organic matter. These animals absorb sulfur in the form of sulfate and then release it as sulfide (S^{2-}). During this process, four oxygen atoms are released, making it a source of oxygen, while the sulfide usually reacts with different metal cations (like iron, yields Pyrite (FeS$_2$)).

Most of the sulfur used as feedstock for the chemical industry is recovered from natural gas and petroleum that contain hydrogen sulfide and other sulfurous organic compounds; however, these chemicals must be removed before the natural gas and petroleum are used. Sulfur-based compounds have different industrial applications: in pharmaceuticals, pigments, and rubber, in which it serves as a cross-linker, forming bridges, in a process termed "sulfur vulcanization," thus controlling the natural or synthetic rubber hardness, elasticity, and mechanical durability. However, ~90%

of the sulfur in industry is used to produce sulfuric acid (H_2SO_4), mainly used to produce phosphoric acid from phosphate during the production of fertilizers.

While sulfur and sulfate are nontoxic, hydrogen sulfide, sulfur dioxide (SO_2), and sulfur trioxide (SO_3), emitted from natural and man-made processes are all toxic. As air pollutants, they affect the respiratory system, particularly the lungs, and may cause eye and skin irritations, whereas hydrogen sulfide is particularly dangerous and can cause death by respiratory paralysis. In addition, both sulfur dioxide and trioxide are constituents of acid rain.

Gold (Au)

Gold is a yellow, shiny, nontoxic, precious metal. It is an excellent conductor of heat, and electricity, and has high resistant to air, heat, moisture, and most solvents. Furthermore, it is the most malleable and ductile of all the metals, and one of the softest and heaviest.

Gold has been known since prehistoric times and was one of the first metals processed by Humanity, mainly because it could be found as relatively pure particles in rivers and streams. It is usually found in various weathered and eroded rocks and minerals that are washed into waterways. In ancient times, gold was considered a perfect substance; as such, for ages, alchemists, philosophers, and pseudoscientists (the predecessors of the scientific field of chemistry) unsuccessfully tried to turn other metals into gold.

Gold vessels, ornaments, jewelry, and coins were used since antiquity, whereas in the twenty-first century most countries keep a stock of gold in a central, national bank, as a guarantee of the value of their money.

The primary reason for gold's very high exchange value is because it is a relatively rare metal (with a global prevalence of less than 0.004 ppm in the Earth's crust), while its processing is relatively easily. Presently, the main hypothesis regarding the origin of gold on Earth is that it came from outer space on meteorites that crashed on our planet. Various studies also show that gold is found in trace amounts in various plants, although gold is not recognized as an essential element for plant metabolism. It seems to be taken up together with other plant nutrients from the soil or water and serves as a stabilizing agent [6]. Gold is also consumed by various microorganisms, bacteria, and fungi that use various metals as energy sources. In addition, although gold is not one of the trace elements in the living body, it is found in the human body in an amount of ~0.003 ppm, close to its prevalence in Nature. The role of gold in the physiology of the human body has not been studied in-depth, but it is known that it plays a role in the maintenance of joint function and the transmission of electrical signals ("nerve impulses") in the nervous system. However, gold mining is one of the most destructive industries in the world, because it uses cyanide (CN^-), a potentially deadly chemical that interferes with the body's ability to use oxygen. Gold mining also displaces communities and contaminates drinking water and soil with toxic metals like mercury (Hg) and lead (Pb).

Mercury is a harmful, potentially fatal, element that causes destructive effects on the nervous, digestive, and immune systems—devastating lungs and kidneys. Lead harms the brain, liver, kidney, and bones, and can severely affect mental and physical development in young children. Thus, gold mining affects many different ecosystems.

Uranium (U)

Uranium is a dense, silvery-white, radioactive metal, which can decay while releasing radiation and other elements. It was used in ancient times in the form of "pitchblende" (now called "uraninite"), a triuranium octoxide (U_3O_8)—the most stable uranium oxide form—to produce yellow ceramic glazes, and during the medieval period, it was also discovered in silver mines. It is both the most-known radioactive substance found in Nature and the heaviest element found in significant quantities on Earth. Uranium has various isotopes, one of which, the isotope $^{238}_{92}U$, which is 99% of the uranium on Earth, is found in tiny concentrations of only a few ppm in soil and water, and is produced from various minerals, such as uraninite, urania, or uranium dioxide (UO_2), and others. The two other uranium isotopes generally found in Nature are $^{234}_{92}U$ and $^{235}_{92}U$, all three of which undergo radioactive decay by the emission of an alpha particle accompanied by weak gamma radiation.

Assessments of the impact of uranium isotopes on humans and other biota found that they have either radiotoxicity or chemotoxicity, whereas chemotoxicity is often of greater significance than radiotoxicity. However, if $^{234}_{92}U$ and $^{235}_{92}U$ concentrations are higher than their natural concentration, or if the progeny of $^{235}_{92}U$ and $^{238}_{92}U$ are present to a significant degree in natural uranium, then the radiotoxicity per unit mass of uranium will be substantially increased, although the chemotoxicity will be essentially unaltered [7].

Naturally occurring uranium is present in the environment, the atmosphere, hydrosphere, lithosphere, and biosphere, at a wide range of concentrations. Its concentration in the environment has sometimes been altered by human activities, mainly mining and its use in nuclear reactors. Its natural, ecological cycle is from terrestrial uplands to the surface drainage network, then downstream to the oceans, eventually incorporated in seabed sediments and recycled to the mantle by subduction processes. As such, uranium also enters freshwater via dissolution and natural processes of erosion of rock and soil. Then, it gains entrance to the "groundwater" (i.e., water located underground, beneath the Earth's surface) via the rivers, etc. In addition, dust particles with uranium originating from unconfined "mine tailings" (i.e., leftover materials after the separation of the valuable fractions from the needless fractions of raw ores) in a terrestrial environment and from "stack emissions" (emanating from smokestacks following a process of incineration) can be windblown into the air and, subsequently, deposited or washed into bodies of water. Presently, the U.S. Environmental Protection Agency (EPA) standards and regulations for uranium exposure in drinking water is 30 µg/L, high above the level in most places. Yet, the movement of uranium originating from anthropogenic sources has greater potential impacts on human health and on the surrounding environment—especially hazardous effluents discharged directly into surface waters and dust from uranium mills, as well as radioactive radiation.

The main use of uranium by Humankind is in atomic reactors that produce nuclear energy and in atomic bombs. In general, 1 kg of $^{235}_{92}U$ has the capacity to produce as much energy as 1500 tons of coal. Beyond the controversial use of atomic bombs, the use of nuclear reactors for the purpose of energy production is also highly controversial, mainly due to the fear of malfunctions or accidents, which have, in the past, and might again, in the foreseeable future, release massive quantities of radioactive radiation and radioactive dust (termed "nuclear fallout"), thereby harming all living entities within the very large circular area surrounding "ground zero" (where the bomb detonates), contaminating its perimeter for many years. Even radioactive waste from reactors or bombs may emit radiation, such that it, too, requires special treatment; radioactive waste byproducts are usually buried deep underground inappropriate lead-lined containers. Nevertheless, the use of nuclear energy has certain environmental advantages, both for the prevention of air pollution and GHG emissions associated with the use of fossil fuels and for conserving of open areas and various metals associated with the use of renewable energies, like solar and wind energies. In fact, today, there are smaller and much safer nuclear reactors, and many believe that the use of nuclear energy is one of the best ways to deal with the increasing consumption of energy while offering decarbonization (phasing out the use of fossil fuels) and causing minimal damage to the natural environment.

References

1. Croswell K (1996). Alchemy of the Heavens. Anchor. ISBN 0-385-47214-5. Archived from the original on 2011-05-13
2. Haynes HM (ed) (2016) CRC handbook of chemistry and physics.97th edn. CRC Press, Boca Raton
3. Araya YN (2005) Hydrosphere. In: Lehr JH, Jack K (eds) Water encyclopedia. Wiley, Hoboken
4. Kuskov O, Kronrod V (2007) Izv Acad Sci USSR Phys Solid Earth 43:42–62
5. https://earthobservatory.nasa.gov/features/CarbonCycle/page1.php. Accessed Jul 2025
6. Cremisini C, Ferrari B, Ghezzi P, Penco S, Roveda L (1999) Gold complexes as therapeutic agents for rheumatoid arthritis. Coord Chem Rev 185–186:457–471
7. Carvalho FP, Fesenko S, Harbottle AR, Lavrova T, Mitchell NG, Payne TE et al (2023) The environmental behaviour of uranium

Chapter 3
The Abiotic Environment

Abstract The inanimate, or abiotic, environment is one of the two main components of Nature. It is primarily composed of inorganic materials like metals, salts, minerals, air, and water, and includes environmental conditions, such as light, humidity, and temperature. Planet Earth's entire environment is divided into several distinct spheres: Geosphere—encompassing all rocky material, molten matter, and minerals found in the Earth's various layers; Lithosphere—consisting of the upper part of the Earth's mantle and the continental and oceanic crusts; Hydrosphere—the aquatic environment that includes all surface and subsurface water bodies on Earth, such as groundwater, oceans, seas, lakes, and rivers, as well as frozen water in the form of glaciers, ice caps, and ice-covered lakes (collectively referred to as the Cryosphere), and even includes the water present within living organisms; Atmosphere—the layer of gases surrounding the Earth, divided into several main layers, based on their composition, electrical properties, and characteristic temperatures. The abiotic environment provides the essential conditions for the existence of life and is vital to the functioning of ecosystems. Changes in this environment, especially due to human activity, can significantly affect the entire ecosystem.

Keywords Abiotic environment · Geosphere · Lithosphere · Hydrosphere · Cryosphere · Atmosphere · Inorganic material · Greenhouse gases · Climate crisis

Immediately after the Big Bang, atoms could not exist, since the Universe was extremely hot and dense, and all matter was in the form of highly ionized plasma. As the Universe expanded and cooled, during the recombination epoch, ions met electrons to form atoms, mostly hydrogen and helium. Afterword, molecules, consisting of more than one atom, either of the same element or of two or more elements, aggregated during the collisions of reactive atoms and ions, thus, instigating chemical evolution.

The first molecule that was formed was helium hydride (HeH^+), a union of helium and one hydrogen proton, though such a creation has never been detected in space until now. This new agglomeration then led to different reactions that further

A. Wolfson, *Chemodiversity and the Ecological Crisis*, SpringerBriefs in
Molecular Science, https://doi.org/10.1007/978-3-032-07363-1_3

yielded lithium hydride (LiH^+) and molecular hydrogen. Later, when elements like carbon, nitrogen, oxygen, and sulfur were released during stellar death explosions, they collided with hydrides and hydrogen molecules, which were the main molecules in the developing atmosphere, thus forming new hydrides, mainly molecules like methane, ammonia, water, and hydrogen sulfide—all having covalent, non-ionic bonds, characterized by sharing electrons in their outermost electron shell. Hydrogen and helium, with traces of these molecules, are common today in the atmospheres of different planets like Jupiter, Saturn, Uranus, and Neptune.

The size of the Universe, which contains all the matter and energy in the stars, galaxies, and the space between them, is infinite. But the "observable Universe," that is, the final part of the Universe that we can observe, is estimated to be a large sphere with a radius of 46.5 billion "light years"—the unit of measure in astronomy that defines the time it takes for light to travel through a vacuum for 1 year. Accordingly, since the speed of light in space is $3 * 10^8$ m^2/s (3 with 8 zeros after it), a light year is $\sim 10^{16}$ m, so the size of the visible Universe is $\sim 46.5 * 10^{25}$ m.

The composition and density of the matter in the Universe may be estimated in several ways, either by summing up the mass of the various objects in the Universe and the matter and energy between them, or by estimating the size of the Universe according to its expansion since the Big Bang. The rate of expansion of the Universe currently indicates that its critical density is $\sim 9 * 10^{-27}$ kg/m^3, which includes matter and energy together. According to various observations, the atoms or "normal" matter constitute only 4.9% of this density, with 26.8% of the matter being "dark matter" and the rest "dark energy" [1].

As stated before, "dark matter" is defined as matter that is not directly visible and does not absorb, emit, or incite light, hence its name, and its existence has been proven by various experiments. The presence of dark energy, however, is even more mysterious and undeciphered. Such energy was defined when various studies showed that the Universe continues to expand at an increasing rate, meaning that the galaxies continue to move away from each other at an increasing rate, contrary to the force of gravity that is supposed to pull them closer together. The prevailing conclusion today is that there is an additional repulsive force between the galaxies, some type of invisible or "dark energy." Finally, the density of the visible matter in the Universe (i.e., the galaxies), which are the largest ordered structures of matter in the Universe that contain millions to billions of stars bound together by gravity and moving together, is $\sim 3 * 10^{-28}$. This is an extremely small number, suggesting a great, empty space between the stars.

As mentioned, the Universe consists of an enormous number of galaxies, having different structures, for example, spiral or elliptical, and containing billions of stars. The Milky Way Galaxy, where our Solar System is located, is an elliptical galaxy, ~ 100 thousand light years in diameter and ~ 1000 light years thick, containing between 100 and 400 billion stars. The mass of the Milky Way Galaxy is $5.8 * 10^{11}$ times the mass of the Sun—a mass that is used as a basic unit of measure in astronomy and is estimated at $\sim 2 * 10^{30}$ kg. Our Solar System, which consists of eight large planets orbiting the Sun, including the Earth, alongside small stars and asteroids, is part of the Milky Way Galaxy. Our Sun occupies 99.86 wt% of our solar system's entire weight, while the Earth, which is ~150 billion kilometers from the Sun, weighs ~6 * 10^{24} kg, and its average diameter is 12,756 km.

Planet Earth was born in the Solar System ~4.6 billion years ago. Like the creation of many other stars, it began as very hot magma, and after of a few hundred million years, the planet began to cool and the heavy elements began sinking into its core (the center of the planet), forming different layers of molten material, while the outermost layer became a solid covering of relatively lighter material. At the same time, oceans of liquid water formed, while Earth was colliding with other small celestial objects. During the Earth's final stage of development, it was bombarded by asteroids that brought other elements to Earth with them. These ongoing, massive processes formed new, solid, liquid, and gaseous compounds that shaped the unique chemodiversity of our world. In fact, the relatively cold temperature in the outermost layer of the geosphere (the lithosphere) is rich with pure solid metals and metal oxides, like silica, lime, and alumina (created during reactions with oxygen), as well as many minerals, composed of different metallic and nonmetallic elements. The ocean, which had formed at a very early stage in Earth's planetary history, was mainly composed of liquid water (the hydrosphere). As soon as this water came into contact with rock, weathering processes began, leaching and dissolving different soluble metal elements out of the rocks (like sodium, calcium, and magnesium), and together with nonmetals from the atmosphere, they yielded different salts, which were soluble in their ionic forms in the hydrosphere and participated as solids in the lithosphere. Finally, the lighter gaseous molecules, especially gaseous nitrogen and oxygen (and their oxides and hydrides, as well as those of carbon and sulfur) were assembled above the lithosphere and hydrosphere, composing our "atmosphere."

Inorganic Materials

The abiotic environment is based on inorganic substances [2]. The inorganic material in Nature consists of several main groups, including certain metallic or nonmetallic elements and a variety of their compounds. Inorganic molecules are usually relatively small and less diverse than organic compounds, and they include the following substances:

Metals Metals, in their neutral form or as components of oxides and minerals, comprise most of the geosphere and its chemodiversity. Metals and their ions have many functions in the biotic environment and serve in many metabolic processes in the living world. Most of these metallic elements are absorbed by plants after being dissolved in water or by "fixation" (i.e., the preservation of biological tissues) with the help of fungi or bacteria. From there, they pass through the food chain to various animals and, ultimately, return to the soil or the water with the decomposition of the dead flora (plants) and fauna (animals).

Acids Generally, "acids" are substances that receive a pair of electrons and, more specifically, lose a proton (H^+) during a chemical reaction or in an aqueous environment, where they form hydronium ions (H_3O^+)—cations created by the addition of single protons to water molecules. By definition, this reaction reduces the pH value below that of the natural pH value of 7, making it "acidic." Strong mineral acids, ranging from hydrochloric acid (HCl) to nitric, sulfuric, and phosphoric acids, play different roles in the natural processes within the abiotic sphere. Such mineral acids regulate the pH of the various bodies of water, thereby assisting in the dissolution of underwater rocks and the minerals within, in the decomposition of substances and in various chemical reactions. Besides strong mineral acids, there are also many organic acids called "carboxylic acids" (that contain a carboxyl group) and have many important functions in living things. Acids are often used in certain man-made synthetic processes and in numerous chemical reactions—either as reagents or as catalysts, also as cleaning and bleaching agents, and even in "electrolytes" (that conduct the electric current) in batteries.

Bases Contrary to acids, "bases" donate a pair of electrons while adsorbing a proton during a chemical reaction that increase the concentration of hydroxide ions (OH^-) when added to water, thus raising the pH value above 7 (above neutral conditions). Moreover, in the case of bases, there are inorganic bases (like ammonia); metal hydroxides (such as: caustic soda; potassium hydroxide (KOH); and sodium hydroxide); and also organic bases (e.g., amines). In addition, like acids, bases also have different functions in natural or synthetic chemical reactions.

Salts Salts are neutral compounds consisting of cations and anions, usually formed in a reaction between an acid and a base, for example, hydrochloric acid reacting with sodium hydroxide, which yields sodium chloride (NaCl, table salt) and water. This ionic bond not only gives these salts a relatively high melting point and a distinct lattice structure, but also high solubility in water, where they dissolve into ions surrounded by the "polar water" molecule or solvent (i.e., a molecule or solvent with unevenly distributed positive and negative electrical charges) by means of a "hydration" process. However, there are also several insoluble salts, whose solubility in water is very low, and in some cases, as in the case of calcium carbonate, they can precipitate and cause scale.

Oxides Oxides are compounds of elements that react with oxygen. A large group of oxides is formed with nonmetals, ranging from water and carbon dioxide or carbon monoxide to gaseous oxides of sulfur and nitrogen. Oxygen is also a central component in various solid oxides, which are very common in the solid matter of the Earth.

Minerals Minerals are solid inorganic substances, mostly crystalline, although amorphous minerals do exist, formed by natural geological processes. There is a very wide variety of minerals in Nature, characterized by different properties, such as different colors, distinct magnetic properties, etc. Their uses are also quite diverse, ranging from building materials (e.g., clay) to precious gems (e.g., rubies).

Composite Materials In Nature, there are also various materials that consist of several molecules not joined together by chemical bonds, rather mixed side by side, such that their aggregation creates new properties, distinct from those of each individual material. For instance, shells that serve as a shelters, protecting mollusks and crustaceans, are built of a combination of calcium carbonate crystals separated by organic matrices of various biopolymers, proteins, and more; also, trunks of trees (main components of biotic environments and used by Humanity as building material or as fuel) primarily consist of polymers, such as: cellulose, hemicellulose, lignin, and pectin, plus oils, resins, water, etc. Another example are spider webs, made of natural composite materials of three types: proteins attached to sugars, known as "glycoproteins"; water; and small molecules that absorb water from the moist environment, enabling the surface of the webs to remain dry and adhere well.

Geosphere

The geosphere includes the Earth itself and all the matter in its various layers, that is, the rocks, minerals, and molten material, and it consists of several key elements. The inner layer (the Earth's core), which is about half the diameter of the Earth and comprises ~32 wt% of the Earth's mass, is divided into a solid inner core under intense pressure and a liquid outer core, assumed to consist mainly of heavy elements, such as iron and nickel. Above the core is the mantle layer, also divided into two sub-layers: the lower mantle and the upper mantle; together with the core, they constitute ~99 vol% of the Earth's entire volume. The mantle is primarily composed of silicon rocks containing oxygen and other elements (i.e., "silicates," with iron and magnesium). Finally, above the core and mantle is the Earth's crust, which is only ~0.5 wt% of the Earth's mass and composes ~1 vol% of the Earth's total volume. It is divided into a marine crust, ~10 km thick and a terrestrial crust between 25 and 75 km thick. The compositions of the marine and terrestrial crusts also differ. While the marine crust consists mainly of silicon and magnesium (as in basalt stones), the terrestrial crust consists of silicon and aluminum (as in granite rock). Nonetheless, the outer layer of the "continental crust" (i.e., the geological continents and shallow seabeds adjacent to their shores, known as "continental shelves") consists of 75% sedimentary rocks, formed from matter derived from older rocks, and 25% from "basic rocks," formed by the crystallization of magma or lava.

The depth on land of the area that consists of the upper part of the upper mantle and the terrestrial and marine crust, also named "lithosphere" is between 40 and 200 km and the seabeds are between 50 and 100 km. The lithosphere mainly contains silica (45.2%) and magnesia (37.6%), and the remainder are oxides of iron (FeO, 8.5%), aluminum (4.5%), calcium (3.6%), sodium (Na_2O) and titanium (TiO_2) [3].

The lithosphere serves as fertile soil, not only supporting various habitats and ecosystems, but also agricultural crops and the establishment of cities and other infrastructures. It is also a major source of a wide variety of minerals, metals, and oxides used in natural and man-made processes.

As human society developed and grew, more and more resources were mined and extracted from the ground, either for direct usage, as in the production of enormous amounts of synthetic compounds, or for indirect usage in various food chains; as such, the quantity, quality, location, and availability of these and other materials being taken from the geosphere are constantly changing. This type of consumption also causes extensive changes in the natural environment and alter the services that natural systems are able to provide to all the species and to humans.

Mining activities typically have a high impact on their surroundings, as well as broader implications for environmental and human health. The consequences of such consumption begin with the designation of mining sites that may cause damage to local habitats and ecosystems (e.g., deforestation). Besides land use impacts, mining also produces a relatively large "water footprint" (i.e., the total amount of freshwater consumed during this production process, since many stages require the use of water). Furthermore, large water footprint also suggests the production of a great quantity of wastewater containing hazardous metals and other elements, usually discharged into local soil or surface and groundwater. Mining generally consumes a large amount of energy and emits air pollutants and GHGs, leaving a very large "carbon footprint," which is the total emission of GHGs from the production process. Additionally, mining usually causes hazardous dust dispersion that may be very harmful to plants, animals, and human beings. Moreover, both pre-mining and mining activities frequently do harm to the local communities, as well.

Hydrosphere

The Earth's "hydrosphere" contains all surface and underground bodies of water, including the oceans, seas, lakes, rivers, and streams; water in the form of glaciers, ice caps, and frozen lakes, also known as the "cryosphere"; groundwater; and even the water within all living things. The hydrosphere weighs $\sim 1.4 * 10^{21}$ kg, which is ~ 0.023 wt% of the Earth's whole weight and covers $\sim 70\%$ of the Earth's surface—$\sim 53\%$ of the northern hemisphere and $\sim 88\%$ of the southern half. Approximately 97 vol% of the water in the hydrosphere is in the oceans, in the form of salty water, while most of the water suitable for drinking ("potable water") is found in glaciers (~ 69 vol%) and groundwater reservoirs (~ 30 vol%); and only 1 vol% of the potable water is found in freshwater lakes and streams (see Table 3.1) [4].

Table 3.1 The distribution of water in the hydrosphere [4]

Location	Volume (10^3 km³)	Part of total volume of hydrosphere (%)	Portion of freshwater (%)	Recycled annual volume (km³)	Renewal period (years)
Ocean	1,338,000	96.5	–	505,000	2500
Groundwater	23,400	1.7	–	16,700	1400
Potable groundwater	10,530	0.76	30.1	–	–
Soil moisture	16.5	0.001	0.05	16,500	1
Glaciers and snow cover	24,064	1.74	68.7	–	–
Water in lakes	176.4	0.013	–	10,376	17
Swamps	11.5	0.0008	0.03	2,294	5
Rivers	2.12	0.0002	0.006	43,000	16 days
Water in biomass	1.12	0.0001	0.003		
Water in the atmosphere	12.9	0.001	0.04	600,000	8 days
Total freshwater	35,029.2	2.53	100	–	–
Total water	1,386,000	100	–	–	–

Water (H₂O)

Water, often called "the liquid of life," is the most common liquid on Earth and the third most abundant molecule in the Universe. It is a small molecule composed of one atom of oxygen that shares a covalent bond with two atoms of hydrogen. These "polar bonds," with a positive charge on the side of the hydrogen atoms and a negative charge on the other side with the oxygen atom, give water its unique properties—especially its ability to dissolve salts while breaking the ionic bonds between the cation and the anion, as well as surrounding and stabilizing both ions. In addition, water can form intermolecular hydrogen bonding—a dipole-dipole attraction between different molecules—characterized by a high boiling point and intramolecular hydrogen bonding with many other organic compounds. Liquid water also has a relatively high "specific heat index" (meaning that it absorbs a lot of heat before it begins to get hot), a characteristic that makes it valuable for industrial purposes. Furthermore, it has very high surface tension, making it sticky and elastic; thus, water forms spherical droplets, rather than spreading out as a thin film and have "capillary action" (i.e., when liquids in a narrow space, like up a straw, will flow in opposition to/or neglecting gravity. Finally, although the term "water" typically refers to the liquid state of this compound, it also exists in a solid state (as ice), or in its gaseous phase (as vapor or steam). Furthermore, under certain conditions, water can also form a "supercritical fluid" (i.e., a state above its critical temperature and critical pressure, in which gases and liquids coexist).

Nonetheless, water is not a vital element of life only in the biotic environment; it also plays a critical role in the abiotic environment, by controlling the temperatures in the hydrosphere and atmosphere—although water is not the factor that ultimately determines their temperatures. The normal temperature on Earth allows most of the water to maintain a liquid form, enabling the presence of life on the planet. The temperature on Earth is determined, first and foremost, by the solar energy coming from the Sun, but the Earth's temperature is also affected by the composition of the atmosphere and the amount of atmospheric water vapor. Atmospheric water vapor that is, in fact, a GHG, is not considered to be a cause of man-made global warming, because it does not persist in the atmosphere for more than a few days while passing through the water cycle. Nevertheless, atmospheric water vapor does amplify the temperature increase caused by other GHGs; this effect is termed the "positive feedback loop"—as the temperature increases, the amount of water vapor in the atmosphere also increases, and vice versa. As such, the addition of more water vapor to the atmosphere may also produce a "negative feedback effect," due to the formation of clouds that deflect sunlight and reduce the amount of energy that reaches the Earth's surface to warm it.

All the bodies of water in the hydrosphere also contain various dissolved salts. The average concentration of salt in the oceans is ~3.5%, that is, 35,000 mg of salt per liter. In contrast, the water in the Dead Sea, which is the second saltiest sea in the world, contains ten times more salt, ~34.2%. The water in lakes and streams contains much less salt, ranging from a few milligrams to hundreds of milligrams per liter, with an average of 70 mg per liter, while rainwater contains ~20 mg of salt per liter.

The most prevalent salt in the hydrosphere is sodium chloride (commonly known as "cooking salt"), along with other salts that mostly contain magnesium, calcium, and potassium ions, together with chlorine, sulfate, and bicarbonate. The list of the significant ions dissolved in seawater, river water, and rainwater, appears in Table 3.2.

Table 3.2 The chemical compositions of the main bodies of water (ppm) [4]

Component	Seawater average	River water average	Rainwater average
Chloride (Cl^-)	19,000	5.75	3.79
Sodium ion (Na^+)	10,500	5.15	1.98
Sulfate (SO_4^{2-})	2700	8.25	0.58
Magnesium ion (Mg^{2+})	1350	3.35	0.27
Calcium ion (Ca^{2+})	410	13.4	0.09
Potassium ion (K^+)	390	1.3	0.30
Bicarbonate ($HCO3^-$)	142	52	0.12
Bromide (Br^-)	67	0.02	–
Strontium ion (Sr^+)	8	0.03	–
Silica (SiO_2)	6.4	10.4	–
Boron (B)	4.5	0.01	–
Fluoride (F^-)	1.3	0.1	–

Besides the various salts, different bodies of water are contaminated by many types of pollutants. While there are natural pollutants (like silt, dust, and animal excrement), the main causes of water pollution, both surface water and groundwater, are man-made actions and processes. Pollutants can come from domestic and industrial sources: various wastewaters and effluents with biochemical contents; direct leakage of hazardous chemical substances (e.g., oil or petroleum from ships); seepage of dangerous substances (like toxic fertilizers and pesticides from agricultural) into the ground, groundwater, and reservoirs; or solid wastes (like plastic items) thrown or washed into streams, rivers, and oceans; and so on. Another type of water pollution is "thermal pollution," which causes a significant increase in water temperature; this occurs when cool water is used as a coolant in power plants and factories and is then returned as hot water to the water source from which it was pumped and negatively effects the ecosystems that exist in that water source.

The pollution of water sources causes wide-ranging, negative health effects. Over the years, water pollution has been one of the main causes of human mortality, mostly resulting from bacterial epidemics (such as cholera and dysentery) transmitted via unclean and untreated water systems. According to U.N. data from 2023, two billion people (26% of the world's population), do not have access to clean, safe drinking water, and 3.6 billion people (46% of the world's population) do not have access to safe sanitation systems. The World Health Organization (WHO) data shows that today more than a million people die each year, due to a water shortage, or the use of contaminated water [5]. For this reason, over the past few decades, various systems for water purification have been developed and there has been much international promotion of standardization and legislation for clean water and safe sanitation.

Water is used by Humankind for many domestic needs: drinking, bathing, and cleaning; agriculture, especially the irrigation of fields and watering the animals; as well as for industrial purposes. Some of the water is absorbed by the processes themselves and the remainder is returned to the environment in the form of wastewater. It is customary to divide the water used by humans into several types:

1. "Blue water"—freshwater originating from surface sources or underground aquifers, used for domestic, agricultural, and industrial needs.
2. "Green water"—water originating from rainwater that has accumulated in sub-surface root-zone soil, which is the primary type of water used for agriculture across the globe.
3. "Desalinated water"—potable water also suitable for agricultural uses, produced from salty water.
4. "Reclaimed water"—treated wastewater purified from various pollutants to certain degrees of cleanliness for reuse in agricultural and industrial processes.

Desalination

The lack of freshwater and the high availability of salty seawater and brackish groundwater have increased the use of water desalination technologies in many parts of the world. During desalination processes, the levels of the salts and the minerals dissolved in seawater (containing $\sim$15,000–50,000 dissolved mg per liter) or in brackish water (with $\sim$1000–15,000 dissolved mg per liter) are reduced to the amount comparable to that found in potable water (less than 1000 dissolved mg per liter). The most common desalination technology today is called "reverse osmosis," which employs a synthetic membrane that serves as a filter, allowing the selective passage of water while rejecting the salt. This process requires a lot of energy to create sufficient pressure (i.e., "osmotic pressure") able to draw the water through the membrane. An older technology is based on the evaporation of the water by heating it and then condensing the clean water vapor, so that the salt remains concentrated in the solution.

Wastewater

Wastewater is water containing various anthropogenic substances, emitted from domestic, commercial, or industrial use. It is customary to divide wastewater into several categories, according to the type and level of pollution they contain:

1. "Black water," also called "effluent"—water that contains a lot of organic matter and bacteria and originates from water that was used in toilets, sinks, and dishwashers ($\sim$40% of the domestic wastewater).
2. "Gray water," also dubbed "dirty water"—water with residues of organic matter but almost no bacteria, originating from baths, washing machines, and air-conditioners ($\sim$45% of domestic wastewater).
3. "Industrial wastewater"—water containing high amounts of chemical, and sometimes also biological, pollutants.

Over the years, the uncontrolled and/or untreated discharge of domestic and industrial wastewater into Nature has caused massive pollution of the water sources, the soil, and the air, which has harmed the Earth's ecosystems, clean water sources, and human health. The understanding that such wastewater may be "recycled," transformed from a hazard into a resource, that it is possible to repeatedly extract useful water from wastewater for its reuse, promotes the development and application of different methods for treating and recycling wastewater for use as irrigation, in industry, and even for drinking. Wastewater treatment entails several stages that determine the final quality of the water and, as a result, the possibilities of its utilization and its value.

The most common treatment is at intensive wastewater treatment facilities via the following stages:

1. "Pretreatment"—removal of coarse or suspended substances.
2. "Primary treatment"—sedimentation and removal of ~30% of the organic matter and ~50% of the suspended solids.
3. "Secondary treatment"—biological treatment that breaks down most of the organic matter, usually using the "activated sludge" method, in which a dense population of bacteria swirls in the treatment pond and breaks down the organic matter.
4. "Tertiary treatment"—removal of materials containing nitrogen or phosphorus and additional filtration of suspended solids.

Another worrisome situation, caused by direct and indirect man-made processes, is the desiccation of rivers and lakes. This can be caused by the excessive exploitation of water resources or the diversion or flood-gating of rivers, or due to climate change, which increases the water evaporation and decreases the amount of precipitation. This desiccation of essential water sources does not only affect the Earth's ecosystems and species, and Humanity's access to potable water, sufficient agriculture, and food security, but it also disrupts certain economic aspects, such as trade and transportation. In fact, recently, there has been a legal, ethico-political movement supporting Nature's own right to water that has made repeated attempts to grant rivers the right to flow and flourish.

Though it may sound ironic, water (the liquid of life) may also be listed among the pollutants. In fact, the emission of water from the ground near roads or houses may increase the humidity, create fog, condense into drops, or freeze, thus changing the local water balance, affecting local ecosystems. Additionally, although it is very difficult to think of water as a pollutant—water may be toxic, similarly to when the electrolytes in the body become imbalanced due to excessive water intake.

Atmosphere

Above the lithosphere and the hydrosphere is the atmosphere, a layer of gases surrounding the Earth. The volumetric composition of the atmosphere on a "dry basis" (excluding the water vapor), is nitrogen (78 vol%), oxygen (21 vol%), and argon (Ar, 0.9 vol%), and small quantities of other gases, such as carbon dioxide, neon, helium, methane, hydrogen, and krypton (Kr) [3].

It is customary to divide the atmosphere into several distinct layers: either according to their specific compositions; or by their electrical properties; or their different temperatures. When distinguishing atmospheric layers by composition, the Earth's atmosphere is divided into two layers. The lowest layer, the "hemisphere" reaches a height of 85 km and contains 99 wt% of the atmosphere, while an upper layer—the "heterosphere" contains the remaining 1 wt%. Dividing the atmospheric layers by their electrical properties describes a neutral layer, up to 50 km, an electrically charged layer, called "the ionosphere," at heights varying between 50 and 1000 km and with various amounts of ionization created by different types of ions (sometimes visible as the stunning "aurora borealis," also known as the "northern lights"). However, the best-known division of the atmosphere is by temperature, and distinguishes five layers, moving from the ground upward, named troposphere, stratosphere, mesosphere, thermosphere, and exosphere, with the two bottom layers being the most significant. The "troposphere" is the first atmospheric layer, closest to the Earth, ~12 km high, containing more air than the other layers and ~80–90 wt% of the matter in the atmosphere. It most affects the weather on the Earth. The second layer, the "stratosphere," is between 12 and 50 km high above the planet, where the air is relatively thin.

Ozone (O₃)

One of the layers in the stratosphere is the "ozone layer," which filters the strong and dangerous UV radiation coming from the Sun to the Earth. This radiation can harm animals and plants and endangers human health; too much UV radiation can, for instance, cause cataracts and certain types of skin cancer, like melanoma.

For almost three decades, since the dawn of the Industrial Revolution, human activity has been causing severe damage to the ozone layer by releasing man-made synthetic chemicals into the atmosphere. Such synthetic chemicals also contain dangerous "halogenated compounds" (i.e., that include fluorine, chlorine, bromine, etc.), from methyl bromide (CH_3Br) previously used as an agricultural pesticide, as well as "freons" or "halons" (i.e., chlorofluorocarbons, CFCs), used in many years in household and industrial air-conditioning systems and aerosol sprays.

After being founded in 1972, in 1978, the "U.N. Environment Program" (UNEP), compiled data on the damage caused by these halogenated compounds to the protective ozone layer. That study was one of the first to show how daily human activity and the use of synthetic substances, although they do not directly and immediately harm public health, can harm the environment and, subsequently, human health. Considering those findings, in 1978, the USA, Canada, and Norway decided to reduce the production and use of sprays that contain CFCs. Yet, it was only in 1984 when it was discovered that the ozone layer was depleting and a large, dark, gaping hole, the size of the continent of Europe, had opened over Antarctica. This sensational discovery raised great concerns and prompted rapid international action that led to the signing of the "Vienna Convention for the Protection of the Ozone Layer" in 1985, and then, in 1987, to the signing of the "Montreal Agreement" (or "Montreal Protocol"), which imposed restrictions on the production and trade of substances that damage the ozone—signed by all U.N. member states. In the following years, that agreement was expanded and dealt with restrictions on trade and the use of other substances that can damage the ozone layer, as well as the implementation of various technologies able to prevent the release of such pollutants into the atmosphere.

The "Montreal Agreement" forced the world to seek good alternatives, for example, replacing the use of spray deodorants with sticks or with roll-ons; replacing highly-toxic freon air-conditioner gas with a newer, eco-friendly refrigerant; and finding eco-friendly solutions for soil disinfection, instead of using a methyl bromide. The quick and effective international action taken in 1987, with the support and participation of the general public, produced tangible results. Observations conducted in recent years show that the Antarctic ozone hole has stabilized and the quantity of CFCs in the atmosphere has decreased. The latest forecasts are optimistic and suggest that, by the year 2050, that ozone hole will have markedly shrunken.

Carbon Dioxide (CO$_2$)

Another important molecule in the atmosphere is carbon dioxide, although its concentration in air is very very low. Carbon dioxide is a colorless, odorless, and harmless molecule found in the air and, in the food and drinks we consume, and it is even produced in our bodies. In 2024, its average concentration in our atmosphere was 422.1 ppm (i.e., 422.1 molecules of carbon dioxide per million molecules of all the gases that constitute the air). This is the highest average annual concentration measured in the atmosphere since direct, official measurements began 66 years ago at the Mauna Loa measuring station in Hawaii, and the highest ever concentration in hundreds of thousands of years (and some claim even millions of years), according to direct and indirect measurements and various simulations.

Is 422.1 ppm a lot or a little? For the sake of comparison: a liter of carbonated drink, which contains a kilogram or 1000 g of water, there are ~6 g of carbon dioxide (6000 ppm). Alternatively, the concentration of carbon dioxide in the blood of humans usually ranges from 840 to 1260 ppb (parts per billion), which is as small as a 300–500 ppb concentration of carbon dioxide in the atmosphere, yet it is able to regulate a person's breathing, by sending a relatively large number of signals to the brain, which responds by activating the diaphragm and drawing fresh air into the lungs. But the answer to the question of whether the concentration of carbon dioxide in the atmosphere is low or high is much more complex, because carbon dioxide plays a decisive role in the existence of all living things on Earth—thus, its concentration in the atmosphere has both direct and indirect impacts on this existence. Carbon dioxide serves as one of the key building blocks of life. It reacts with water in the presence of sunlight causing photosynthesis that produces "glucose," a sugar necessary for the production of energy in biosystems, and which also serves as a building material in the creation of polysaccharides, such as the essential biopolymer, cellulose. Moreover, the process of photosynthesis is accompanied by the release of oxygen into the atmosphere, which is also an essential element for the existence of life.

Carbon dioxide is also a GHG, and together with other GHGs it is essential for keeping solar heat close to the Earth, thus maintaining life-supporting temperatures. Together, these GHGs absorb the heat and infrared (IR) radiation energy deflected from the Earth following bombardment by strong solar radiation. Since a large portion of the solar radiation is not absorbed by the soil and oceans (due to their lower temperatures), it is deflected into the atmosphere at a longer wavelength, as IR, some of which is absorbed by the atmospheric GHGs, while the rest escapes the atmosphere. This energy is absorbed when the IR radiation causes the chemical bonds between the atoms of the GHG molecules to vibrate and rotate. Then, this energy is released by the GHGs back into the atmosphere and absorbed by many other GHG molecules. In this way, the GHGs create a kind of heat-containing atmospheric blanket (i.e., keeping the deflected heat close to the surface of the Earth).

Over time, there has been a continuous increase in the concentration of GHGs (including carbon dioxide) in our atmosphere, causing the Earth's mantle to thicken, thus less radiation is being deflected into outer space, while the amount of heat trapped near the Earth's surface continues to increase. This phenomenon causes the troposphere to warm up, like in a greenhouse, so it is called the "greenhouse effect." The ever-increasing amount of heat trapped near the Earth's surface is also causing a constant increase in the average temperature on the Earth's surface, the phenomenon known as "global warming," as well as significant long-term changes in the Earth's weather and climate, known as "climate change." Furthermore, the changes in the atmospheric GHG concentration and the Earth's surface temperature are causing a variety of chemical, physical, and biological changes in our atmosphere, hydrosphere, lithosphere, and biosphere—altogether threatening the processes that exist on the Earth, in the natural environment, and in the human environment— referred to as the "climate crisis."

As mentioned before, at the molecular level, the climate crisis is manifested as a steady increase in the concentration of GHGs in the atmosphere, primarily indicated by the concentration of atmospheric carbon dioxide. Direct measurements show that the concentration of carbon dioxide increases steadily over time, breaking previous records annually. Nevertheless, some studies, based on geological findings and simulations, claim that the concentration of carbon dioxide in the atmosphere $\sim$50 million years ago was over 1000 ppm [6], and that concentrations like those being measured today already existed in the atmosphere $\sim$10–15 million years ago, while, theoretically, concentrations of $\sim$400 ppm of atmospheric carbon dioxide would have been found 3.4–3.6 million years ago [7]. However, more precise measurements of carbon dioxide, taken from air bubbles trapped in Antarctic glaciers, indicate average concentrations of 270 ppm over the past 800,000 years (concentrations cycling between 180 and 330 ppm) [8]. Direct measurements show that, since 1958, the concentration of carbon dioxide has been increasing every year by $\sim$2 ppm, reaching 422.2 ppm by the end of 2024. A special report had stated that 350 ppm is the maximum safe concentration of carbon dioxide in the atmosphere, which would set the increase in global temperature at 1 °C above the pre-industrial period till the end of the current century, thus preventing climate stability from being undermined [9]. A similar number was also determined as part of a "planetary boundaries model" [10]. At the same time, certain forecasts showed that to maintain no more than an increase of 2 °C in average temperature, relative to the pre-industrial period, the concentrations of carbon dioxide in the atmosphere should be below 450 ppm.

Though natural changes in the atmospheric concentration of carbon dioxide and the corresponding changes in the Earth's surface temperature occur all the time, the frequent, often extreme, climate changes we are experiencing now are caused by the ongoing, man-made increase in the atmospheric concentration of carbon dioxide. There is no doubt that many man-made processes operating over several centuries have inflated the emissions of carbon dioxide and other GHGs, whether resulting from the combustion of fossil fuels or due to various agricultural and industrial processes.

Indeed, the increase in carbon dioxide emissions has continued steadily. If, in 1900, those global emissions stood at 1.95 billion tons and, at the start of the twenty-first century, they had risen to 25.23 billion tons—by 2019, they had reached an unprecedented peak of 36.7 billion tons (an increase of ~1780% since 1900 and ~46% since the beginning the current century). In addition, about half the amount of carbon dioxide emitted by such combustion processes have dissolved in the oceans, while the other half remain in the atmosphere, raising the concentration of atmospheric carbon dioxide to 422.1 ppm in 2024. According to a registry of the U.S. National Oceanic and Atmospheric Administration (NOAA), since 1880, the temperature on Earth has risen by 0.08 °C every decade; meanwhile, over the past 40 years, the rate of increase has gotten faster, reaching the rate of 0.18 °C per decade. Accordingly, the average temperature on the surface of the Earth, while accounting for the temperatures on land and sea, in summer and winter, and by day and night—has increased by 0.98 °C since 1900 and by 1.19 °C since the pre-industrial period. Moreover, the 10 warmest years measured (since measurements began) occurred during the past 15 years. Note, however, that an increase in the global average temperature does not reflect local or spot changes in the temperatures of specific places across the globe. A peak temperature may be measured in summer or winter, by day or night, or over a longer duration in some specific location that is more than 2 °C higher and even more than 5 °C higher than the local average temperature in that same place, for the same time.

Not Just Carbon Dioxide

Carbon dioxide is indeed the primary GHG emitted by man-made processes, but it is not the only one, nor is it the most aggressive. As mentioned before, water, which evaporates from the oceans and is emitted by natural processes, is undoubtedly and without any competition the most significant GHG in the atmosphere, with a contribution of ~60% to the natural effect of keeping heat in the atmosphere. But the amount of water vapor in the atmosphere does not change due to direct emissions resulting from human processes. Nonetheless, a rise in temperature is caused by other man-made GHG emissions that indirectly affect the evaporation of water from the oceans. Unlike other GHGs, water vapor does not accumulate in the atmosphere, nor does it remain there for any length of time.

Except for carbon dioxide, which makes up ~76% of all the GHGs emitted by human processes, the main anthropogenic GHGs are methane (16%), nitrous oxide (6%), and the rest (2%) are various gases that contain fluorine, among them:

Table 3.3 Global warming potentials (GWPs) of greenhouse gases (GHGs)

Gas type	Relative quantity (%)	Lifetime (years)	GWP for 20 years	GWP for 100 years
CO_2	76	300<	1	1
CH_4	16	12	86	28
N_2O	6	121	264	265
CCl_2F_2 (CFC-12)	2	100	10,800	10,200
$CHClF_2$ (HCFC-22)		12	5280	1760
CF_4		50,000	4880	6630
SF_6		3200	17,500	23,500

HFCs—hydrofluorocarbons; PFCs—perfluorocarbons; and sulfur hexafluoride (SF_6) (see Table 3.3). The various GHGs have different effects on global warming. Thus, it is customary to use their GWPs and the times it takes them to deteriorate in the atmosphere. Thus, for example, 1 kg of methane is equivalent to 82 kg of carbon dioxide, in terms of its impact on global warming over a period of 20 years. However, since methane breaks down relatively quickly in the atmosphere (within ~16 years), compared to the several hundred years it takes for carbon dioxide to deteriorate, this potential changes over the years. For instance, over a span of 100 years, the impact of 1 kg of methane will have become equivalent to that of only 28 kg of carbon dioxide. As such, to summarize the total carbon emission for different types of discharged GHGs, their "carbon dioxide equivalents" (CO_2eq) are provided in Table 3.3, for relative emission quantities at three different time intervals.

Methane (CH_4)

Methane, a colorless and odorless gas that occurs abundantly in Nature, is the simplest hydrocarbon, composed of a carbon atom and four hydrogen atoms. As mentioned previously, methane was produced on Earth at a relatively early stage. About 3.5 billion years ago, the concentration of methane in the Earth's atmosphere was 1000 times as much than it is now, and it was mostly released into the atmosphere by volcanic activity. Moreover, at that time, ancient bacteria (some of Earth's earliest lifeforms) converted hydrogen and carbon dioxide into methane and water. Oxygen would not become a major component of the atmosphere until after photosynthetic organisms evolved. At that time the methane concentration in the air was drastically reduced, due to its oxidation by gaseous oxygen, yielding carbon dioxide and water.

Global average methane concentrations in the atmosphere reached ~1932 ppb at the end of 2023, more than two-and-a-half times pre-industrial levels [11].

Major natural sources of methane include emissions from wetlands and oceans, and from the digestive processes of termites; however, most of the methane is released into the atmosphere by man-made processes, making it the second largest anthropogenic GHG in the atmosphere. As the primary component of natural gas, methane is emitted during the processing of natural gas: production, separation, conduction, and storage, as well as the production of coal and petroleum. In total, these processes are responsible for the emission of ~25% of the methane emitted. However, most of the methane is released on Earth during the digestion of organic matter by cattle and sheep, alongside anaerobic fermentation (29% of the methane); the fertilization of agricultural crops (14%); the decomposition of rotting organic waste (11%); and the decomposition of wastewater (9%).

Nitrous Oxide (N$_2$O)

The third largest anthropogenic GHG is nitrous oxide. It is a more aggressive GHG than carbon dioxide or methane and, over the course of 100 years, a single molecule of nitrous oxide has caused 300 times more global warming than an individual molecule of carbon dioxide. In fact, although the calculated share of nitrous oxide among all the GHGs is only 6 wt%, regarding its impact on the climate (during a 100-year period), it is responsible for ~70% of the global warming, while carbon dioxide can only be blamed for causing ~3% of it. In addition, nitrous oxide reacts destructively with ozone gas, damaging the ozone layer. Ironically, nitrous oxide is very stable in the atmosphere and does not decompose for ~120 years.

Nitrous oxide is a nontoxic chemical compound, mainly used as an anesthetic or numbing agent, during surgeries or dental treatments. It is also used as a food additive that promotes foaming, and as a fuel additive in racing cars and motorcycles, immediately improving combustion and increasing engine output. One major source of nitrous oxide emission is from the natural decomposition of nitrogen compounds in plants and animals; however, its man-made emissions are primarily from the decomposition of fertilizers that contain nitrogen compounds and manure, resulting from the expansion and intensification of agriculture. The use of fertilizers has many benefits, primarily for the provision of enough food to feed the world's population, which is growing exponentially year by year. Furthermore, the increase in the quantity of cattle and sheep "feed crops" (e.g., oats and alfalfa) have led to a sizable addition in the utilization of nitrous oxide and to the considerable contribution of organic nitrogen compounds, discharged within the soil either in the form of urea, $CO(NH_2)_2$ (coming from urine), or from the deterioration of proteins in animal losses. Alongside these, there are also industrial emissions, resulting from the combustion of fossil fuels containing nitrogen compounds (although it is important not to confuse nitrous oxide (N_2O) with nitrogen dioxide (NO_2)—which is the main nitrogen compound emitted by burning processes).

Another jump in emissions of nitrous oxide was recorded with the intensification and expansion of agriculture in China, India, and Brazil over the last 20 years, alongside the increasing use of biomass both as an energy source and as a raw material for the chemical industry (replacing fossil fuels). In fact, nitrous oxide threatens to become yet another major nemesis of life on Earth; therefore, its emissions must be reduced, despite the constant increase in the world's population and the growing extent of the agricultural crops. Various measurements recently presented indicate that global nitrous oxide emissions have been steadily increasing over the past 120 years and while, in 1900, the atmospheric concentration of nitrous oxide was ~270 ppm—today, its concentration has reached ~340 ppm [12], meaning an increase of more than 25%. Moreover, it was recently discovered that most estimates regarding the amounts of gas emissions and the rise in nitrous oxide emissions were underestimates.

The key means for dealing with nitrous oxide emissions are by the intelligent and efficient use of fertilizers and by including special additives to fertilizers that inhibit the formation of nitrous oxide in the soil. In addition, proper watering and the prevention of excess moisture in the soil, which activates the microbes that break down the fertilizer to flourish; proper mixing of the soil; and growing crops with better nitrogen fixation—all these regulate emissions.

Climate Crisis

As mentioned above, altering the chemical composition of the atmosphere by increasing anthropogenic GHG emissions has led to various acute changes. Physical changes in the temperature, the speed of underwater currents, and the "wind regime" (i.e., average velocity, Weibull probability distribution, and turbulence) may promote heat and cold waves, storms, flooding, and fires. These changes are also manifested by the melting of glaciers at the poles, the rising sea levels, the development of "urban heat islands" (i.e., area in cities that are hotter than their surrounding areas), the concentration of oxygen in the oceans and their acidification, etc. Such climate changes are not theoretical science; they are happening here and now, and their effects are clear and tangible. As a result, the climate crisis has a massive and dramatic effect on the natural environment, mostly expressed as damage to natural habitats, the extinction of many species, and the disruption of ecosystems, as well as having far-reaching socioeconomic effects. Climate change causes the shrinking and desiccation of streams, lakes, and various water sources (e.g., the accelerated shrinkage of Utah Lake in the USA in recent years) and leads to droughts that damage the food security of millions of people (e.g., the increasing frequency and intensity of droughts and starvation in Africa).

This climate crisis damages property and paralyzes essential infrastructures, ranging from electricity and water systems to transportation systems—disrupting our orderly lives. It affects the economy, namely the prices of goods, the cost of living, trade agreements, and the economic performance and competitiveness of farms, to name just a few examples. Furthermore, its consequences cause greater, even extreme, socioeconomic disparities within and between countries, such as growing inequality, poverty, hunger, epidemics—ultimately leading to conflicts over natural resources and regarding refugee traffic. It will no doubt claim the lives of multitudes of people, directly and indirectly. All these climate changes, which are becoming more frequent, longer lasting, and extreme, across the globe, together define a major crisis, the climate crisis, which is currently the biggest threat to our planet and Humanity.

Air Pollution

In addition to GHGs, human activities have introduced many other molecules into the atmosphere, most of which are harmful and toxic. Air pollution is emitted by a wide variety of processes conducted in energy production, transportation, agriculture, industry, waste treatment, the domestic sphere, and elsewhere. It is customary to divide air pollutants into "primary pollutants," those emitted directly by various processes, and "secondary pollutants," created in the atmosphere by certain reactions between primary pollutants or with other components in the air.

The primary man-made air pollutants are organic substances: oxides of nitrogen (NO_X) and sulfur (SO_X); carbon monoxide; heavy metals; and particles emitted by combustion processes—"irrespirable particles" with diameters greater than 10 μm (>PM10); "respirable particles" that penetrate the lungs, with diameters of less than 10 μm (<PM10); and "fine respirable particles" that are less than 2.5 μm (<PM2.5) and can also penetrate into the blood and brain—classified as "carcinogenic" (i.e., cancer causing). Among the secondary pollutants, it is customary to list particles that are formed in atmospheric reactions and "bad ozone," that is, ozone at ground level rather than high atmospheric levels, since it is a strong oxidizing agent.

According to the WHO, certain exposure to air pollution currently leads to the annual death of ~7 billion people (~4.2 million of them from outdoor pollution and the rest from indoor pollution) [13], although there are also higher estimates. In 2019, the WHO also stated that ~99% of the world's population reside in places where the air-quality levels recommended by the WHO are not being met.

Outdoor air pollution is caused by vehicular traffic, especially that of heavy vehicles, and by power plants and industrial factories; such pollution is the number one environmental cause of mortality and morbidity in urban areas. According to the Organization for Economic Cooperation and Development (OECD), by the year 2060 outdoor air pollution will claim the lives of ~6–9 million people [13]. Mortality from air pollution is mainly due to heart, blood, and various respiratory diseases, and there is also widespread morbidity from cancer and other environmental ailments. Moreover, air pollution has extensive personal and socioeconomic consequences, due to health costs; loss of income; and loss of manpower and productivity. Furthermore, morbidity and mortality are not evenly distributed between the different regions of the world, within each country, and in cities, nor between different segments of society. In this respect, the weaker social strata often suffer more from the air pollution, for which they, in many cases, are less responsible. Those at the bottom of the social hierarchy have fewer means for fighting against pollutants and treating diseases, as well as having less access to supportive, high-quality health services.

References

1. Adam R (2022) Dark matter. Britannica
2. Cotton F, Wilkinson AJ, Gaus PL (1995) Basic inorganic chemistry. Wiley, Hoboken
3. Haynes HM (ed) (2016) CRC handbook of chemistry and physics.97th edn. CRC Press, Boca Raton
4. Araya YN (2005) Hydrosphere. In: Lehr JH, Jack K (eds) Water encyclopedia. Wiley, Hoboken
5. IPCC (2014) Synthesis report in climate change 2014: contribution of working groups I, II and III to the fifth assessment report of the Intergovernmental Panel on Climate Change. IPCC, Geneva
6. Inglis GN, Bragg F, Burls NJ, Cramwinckel MJ, Evans D, Foster GL et al (2020) Chem Geol 211:47–69
7. Brigham-Grette J, Melles M, Minyuk P, Andreev A, Tarasov P, DeConto R et al (2013) Science 340:1421–1427
8. Bereiter B, Eggleston S, Schmitt J, Nehrbass-Ahles C, Stocker TF, Fischer H et al (2015) Geoph Res Lett 42:542–549
9. Hansen J, Kharecha P, Sato M, Masson-Delmotte V, Ackerman F, Beerling DJ et al (2013) PLoS One 8:e81648
10. Rockström J, Steffen W, Noone K, Persson Å, Chapin III FS, Lambin E et al (2009) Ecol Soc 14:32
11. He J, Naik V, Horowitz LW, Dlugokencky E, Thoning K (2020) Atmos Chem Phys 20:805–827
12. Hall BD, Dutton GS, Elkins JW (2007) J Geophys Res Atmos 112:1–9
13. WHO (2022) Fact Sheets. https://www.who.int/news-room/fact-sheets/detail/ambient-(outdoor)-air-quality-and-health

Chapter 4
The Biotic Environment

Abstract "Life" is the defining feature of Planet Earth, and "biodiversity"—the variety of all living organisms on Earth—is the key characteristic of life itself. Biodiversity, which serves as an indicator of the stability of the natural environment, consists of a vast range of organic and inorganic compounds and forms the ecosystems that make up the Biosphere (also called the Ecosphere). This diversity is steadily declining and constantly under threat due to human activities—primarily hunting, habitat destruction for the construction of human settlements and infrastructure, agricultural development, the emission of air pollutants, sewage and waste discharge, climate change, and the spread of invasive species. Since the "Industrial Revolution," and especially since the 1970s, an accelerated mass extinction of numerous species has been occurring. This event, often referred to as the "Sixth Extinction" event, is the first, modern mass extinction caused by human actions. In addition to the species already lost, countless others (including *Homo sapiens*) are under threat, and many ecosystems have undergone profound (even irreparable) transformations. Altogether, these developments are contributing to a crisis marked by the loss of biological richness, which is becoming a dominant factor threatening the continued existence of Humanity.

Keywords Biotic environment · Biodiversity · Organic compounds · Biosphere · "Sixth Extinction"

The next stage in the evolutionary process, after "atomic evolution," was that of "chemical evolution" (also known as "abiogenesis" or "prebiotic evolution")—the stage preceding the formation of biological matter. It was characterized by the creation of the first organic molecules, molecules consisting of a skeleton of the carbon and hydrogen atoms that constitute the unique component of our emanant world [1]. Although there are different, even controversial, hypotheses as to how these molecules were formed, clearly, they constituted the basis for the creation of diverse, basic bioorganic compounds, like the amino acids that later became

© The Author(s), under exclusive license to Springer Nature Switzerland AG 2025

A. Wolfson, *Chemodiversity and the Ecological Crisis*, SpringerBriefs in Molecular Science, https://doi.org/10.1007/978-3-032-07363-1_4

components of proteins and sugars, which, in turn, became the building blocks of carbohydrates and nucleotides, from which, eventually, among other things, nucleic acids were formed.

At this stage, molecular evolution began to take place, building and changing the compositions and sequences of cellular molecules, such as proteins and genes. This sequence of evolutionary processes—atomic evolution, chemical evolution, and molecular evolution—completed the creation of chemodiversity.

Just as the first organic matter was created from inorganic matter, the first living matter in the Universe—matter that can reproduce, sustain itself, and undergo constant change—was created from inanimate matter ~3.5 billion years ago, by means of a process known as "abiogenesis." This process initially produced the microorganisms: bacteria, tiny algae ("microalgae"), and small fungi ("yeasts"). Then, ~542 million years ago, plants and animals appeared, and "biological evolution" began. The key factor in biological evolution is the process of "natural selection," i.e., the genetic change and adaptation of species to changing environmental conditions, ultimately causing the appearance of better adapted species and disappearance of maladjusted species. Some scientists claim that ~99% of the species on Earth that have become extinct thus far died out due to those continuous evolutionary processes [2]. Natural selection defines the living ("biotic") environment on Earth, as well as the diversity of species ("biodiversity") in Nature—referring to the total number of species and genes and, similarly, to their variations and the interrelationships between them.

Organic Compounds

As mentioned above, organic compounds have structures consisting of carbon and hydrogen atoms linked to each other, and to which other elements, such as oxygen, nitrogen, phosphorus, sulfur, and others are sometimes added [3]. The variety of organic substances in the world is very large, ranging from small molecules to huge molecules, which provides the variety and diversity of the components in the biotic world; but, as previously stated, this is also true regarding the many man-made processes, constantly producing various, diverse synthetic organic compounds. These natural and synthetic organic materials contain various functional groups, groups that have a marked effect on the physical and chemical properties of each material, for example, melting and boiling points, solubility, and chemical reactivity.

Hydrocarbons

Hydrocarbons are organic substances that contain only carbon and hydrogen atoms. For instance, in the case of an "alkane" with "saturated bonds," its carbon atom forms a bond with two other carbon atoms and with two hydrogen atoms in a "covalent bond" (i.e., when two atoms share an electron pair). There are also alkenes and alkynes with "unsaturated bonds," having double or triple bonds between carbon atoms, respectively (i.e., bonds in which the carbon atoms share two pairs or three pairs of electrons, respectively). Hydrocarbon also have various "isomers" (i.e., molecules having the same number of carbon and hydrogen atoms—thus, sharing the same chemical formula, but having different chemical structures and properties). One such isomer is butane, consisting of four atoms of carbon and 10 atoms of hydrogen; its chemical formula is C_4H_{10}, which can appear either in the form of a straight or a branched chain. Two more examples are "normal-butane" ("n-butane"), which contains four carbons linked together in an "unbranched" chain ($CH_3CH_2CH_2CH_3$), whereas "isobutane" has a "branched" chain ($CH(CH_3)_3$). Some of these hydrocarbons are described as being "aliphatic" (i.e., with linear open chains) and can form regular, double, or triple bonds, and others can also form ring structures. Moreover, hydrocarbons can also form "aromatic" substances (i.e., containing sigma bonds and delocalized pi electrons between carbon atoms in a ring), meaning that there are rings consisting of C–C bonds, located between a single bond and a double bond, in which the electrons revolve around the entire molecule and not only around the two bonded atoms (giving them their description as "unlocalized electrons"), as found in benzene (C_6H_6). All these hydrocarbons are used as fuels and basic materials to produce a wide variety of chemicals having additional functional groups.

Benzene

Benzene, the simplest aromatic compound, a colorless liquid with a sweet odor, is also a highly flammable molecule with very low solubility in water. It is created during natural processes, mainly by volcanic activity and forest fires, and eventually becomes an essential component in the formation of petroleum; human activities create benzene by burning wood or fossil fuels, and also produce it in cigarette smoke and during certain industrial processes.

Benzene has had several uses since it was discovered. It was used to decaffeinate coffee and as a significant component in many consumer products, ranging from paint solvents to rubber cements and spot removers, and in various dyes, glows, and other materials. Yet, with the understanding that it causes severe health effects, it was replaced by other compounds, many of them aromatics, and its solo use as a product was reduced. Today, benzene is mostly used to produce other commercial aromatic compounds, half of it for ethylbenzene, followed by styrene (the raw material for the production of polystyrene, PS), and the rest for the preparation of phenol, aniline, and dodecylbenzene for dyes, detergents, and commodities.

Benzene is commonly found in the environment due to man-made activities, from fuel production, conduction and storage leaks, and various industrial processes that release benzene into air, water, and soil. In addition, benzene is emitted into the atmosphere by the burning of fuels in power stations and motor vehicles. As such, everyone is exposed to a small amount of benzene from outdoor air every day. Yet, everyone is also exposed to benzene indoors, at even higher concentrations than outside—and not just from polluted air entering our homes, but also because benzene is released from certain in-house sources, ranging from cigarette smoke to paints, carpets, furniture wax, and detergents. As such, motor exhaust and industrial emissions account for only ~20% of our total exposure to benzene. In fact, in developed countries, about half the exposure to benzene results from exposure to tobacco smoke. In underdeveloped countries, and especially in places where open fires are used to warm the house and cook, the indoor benzene levels are much higher, causing higher morbidity and mortality. Furthermore, we are also exposed to benzene through our food, beverages, and drinking water. Where the outdoor airborne benzene levels range from 0.02 to 34 ppb, its level in the drinking water is typically less than 0.1 ppb.

Exposure to benzene can cause eye and skin irritation, skin tissue injury, headaches, drowsiness, and dizziness. Benzene can enter the human body through our lungs, our skin and, in trace amounts, via gastrointestinal processes, thus entering the bloodstream and damaging the body's cells. Its negative health effects depend on the amount and length of time of the exposure, as well as the age and medical condition of the exposed person. While acute exposure to benzene can cause death, long-term exposure over years mainly effects the blood cells and causes harmful effects in the tissues that form blood cells, especially damaging the bone marrow. Benzene can also harm the immune and nerve systems and may be harmful to the reproductive organs. Finally, benzene is a "carcinogenic," toxic to humans, and may cause different types of blood cancer (e.g., leukemia).

Organic Compounds with Halogens

Organic compounds that contain various halogens (e.g., chlorine or fluorine) are termed "haloalkanes," "alkynes," or "alkyl halides." These substances are used in many chemical reactions, in which the halogen is replaced by other functional groups, and, in some cases, they are used as refrigerants (like halons or freons), some of which have recently been removed, because they damage the ozone layer. In addition, halogenated hydrocarbons are used as disinfectants and flame retardants.

Organic Compounds with Oxygen

"Alcohols" are a group of organic substances with one or more hydroxyl bonds, consisting of oxygen bonded to hydrogen (-OH) that is further bonded to carbon. Alcohol molecules have relatively high polarity and high solubility in water, but also relatively low volatility and a high boiling point. The most familiar alcohol is "ethanol" or "ethyl alcohol" (C_2H_5OH), which is found in alcoholic beverages and has applications as an antiseptic and an agent for combustion. Ethanol is naturally produced by the fermentation of fruit and grain sugars, by yeast or bacteria to produce energy. Other alcohols are produced by the decomposition of biomass, separated by distillation or other fossil-fuel separation processes, or synthetically created from hydrocarbons or alkyl halides in laboratories. They are used as solvents in chemical reactions, paints and other formulations, and in a variety of cosmetic products, cleaning agents, etc.

Some alcohols have more than one hydroxyl groups, such as "ethylene glycol" ($CH_2(OH)CH_2OH$), which has two hydroxyl groups and is used as an antifreeze; or "glycerol" ($CH_2(OH)CH_2(OH)CH_2OH$), which has three hydroxyl groups in the same molecule, and originates from the breakdown of fats and oils called "triglycerides" and is used in cosmetics, the production of soaps, etc.

Apart from alcohols, there are two other functional groups that have shared bonds between carbon and oxygen—"ketones" and "aldehydes"—in which the oxygen is linked to the carbon in a double bond (C=O), called a "carbonyl group." In ketones, the carbonyl group is bound to two carbons on both sides (e.g., in acetone ($CH_3(CO)CH_3$)), while in aldehydes the carbonyl is bound to carbon and hydrogen (like hexanal ($C_5H_{11}(CO)H$)). Natural ketones and aldehydes are found in fruits and used as flavoring agents; their synthetic versions are produced from alcohols and applied to the production of various other substances.

Another group in which there are carbons and oxygens is "carboxylic acid" (COOH), in which a carbonyl group is attached to a hydroxyl group, as in acetic acid (CH_3COOH) (the active ingredient in vinegar) or lactic acid ($CH_3CH(OH)COOH$). Natural and synthetic carboxylic acids are used as preservatives and antimicrobial agents in food, medicine, and perfumes, as well as in the production of polymers. Finally, a reaction between a carboxylic acid and an alcohol produces an "ester" (-COO-), and various natural esters are used as fragrances, while synthetic esters are used in perfumes, but also as solvents and raw materials for the production of polymers, called "polyesters."

Organic Compounds with Nitrogen

Another group of organic substances contains nitrogen bound to carbon, mainly including amines, amides, nitro groups, and a nitrile group. "Amines" are organic bases, in which nitrogen is bonded either to one carbon atom and two hydrogen atoms ("primary amine"), or to two carbons and one hydrogen ("secondary amine"), or to three carbons ("tertiary amine"). In Nature, for the most part, they are components of amino acids—the building blocks of proteins—though they also have diverse synthetic applications in medicines, dyes, and polymer production. The "amides" (-CONH-) are created by reactions between a carboxylic acid and an amine (as in the case of esters), and form the basic bond that builds proteins, known as a "peptide bond." A "nitro group" has a nitrogen atom bound to two oxides (-NO_2) and to one carbon atom. Natural nitro compounds are very rare, while its synthetic ones are mainly used in the production of explosives, such as trinitroglycerin (first used in a dynamite explosive invented by Alfred Nobel in 1867) or produced as a "reaction intermediate" (i.e., *a molecular entity yielded by an early stage in a stepwise chemical reaction, later consumed during that same reaction)* or during the production of pharmaceuticals or agricultural chemicals. The "nitrile" is a group with a triple bond of carbon and nitrogen (-CN), found, for example, in cyanide or in various organic solvents. Finally, when an atom of nitrogen replaces one of the carbon atoms in an aromatic ring, the resulting substance is called "pyridine," which is used as a building material in various chemical reactions.

Organic Compounds with Phosphorus and Sulfur

Natural organic molecules containing phosphorus appear in Nature in the "phospholipids" (i.e., fatty acids containing a phosphate group) that compose cell membranes. The synthetic phospholipids used in the production of pesticides are often considered very toxic. Sulfur is usually found in organic compounds or bound to two additional carbon atoms, replacing a carbon atom, as in an amino acid of the methionine type, or when bound to one carbon atom and one hydrogen atom, a structure termed a "thiol": as in the amino acid, "cysteine"; or in the "mercaptans" used as strong odorants; or in the compound "ethyl mercaptan" (CH_3CH_2SH), added to odorless cooking gas to enable the detection of gas leaks.

Finally, there are natural organic substances that have several different functional groups in the same molecule, such as amino acids, which carry a functional carboxylic acid group and a functional amine group. There are also synthetic organic substances, such as paracetamol or acetaminophen (commonly known by its trade names as acamol, novimol, paramol, etc., and used as an analgesic and antipyretic), which contains an aromatic ring with a hydroxyl bond and an amide group on it.

Vital Natural Organic Compounds

All living elements in Nature cannot exist without four groups of essential organic molecules: nucleic acids, proteins, carbohydrates, and lipids (Table 4.1).

The main organic material in Nature is "cellulose," the key component that reinforces plant cell walls, accounting for about a third of the plant's weight. In addition to cellulose, there is a large variety of sugars, such as glucose, fructose, and "disaccharides" (like "sucrose," the sugar often used to sweeten processed foods and drinks), and additional "polysaccharides" (e.g., starch, "chitin"—the most abundant aminopolysaccharide polymer occurring in Nature), and others. In the biotic world, these compounds serve as nutrients, sources of energy, and building materials.

Proteins are chains of amino acids, key components in all organisms, responsible for many processes in plants and animals, among them the regulation of the metabolic processes, the water balance, the transport of food and oxygen in the blood, and the building of new tissues. The main lipids are vegetable oil and animal fat, termed "triglycerides," consisting of three linked (saturated or unsaturated) fatty acids. These components serve as natural energy sources, insulating materials, and

Table 4.1 Key natural organic substances

Component	Structure	Main function	Examples
Nucleic acids	Long polymers consisting of a nitrogen-containing aromatic base, attached to a pentose, five-carbon sugar, and a phosphate	Transmission of hereditary material and the construction of proteins	DNA and RNA
Proteins	Large and complex compounds made of folded chains of amino acids	Building cells and cell organelles, regulating chemical and physical processes in animals and plants	Enzymes and collagen
Carbohydrates	From sugars to polysaccharides	Short-term energy production and building material in the cells	Glucose, fructose, starch, and cellulose
Lipids	Small molecules, usually with a hydrophilic head and a hydrophobic tail	Long-term energy production and structural material in cell membranes	Fat, wax, and steroids

as the building blocks for creating cells. Humanity uses oils for a variety of very common purposes, ranging from cooking (especially frying, seasoning, and to absorb vitamins) to oils used in cosmetics and for lighting, which are extracted from seeds, nuts, or legumes, and even from animals (like fish oil). Organic oils are also used as lubricants for machinery, usually in mixtures with various minerals and soaps (e.g., grease), called "industrial oils." In the past, salts produced from free fatty acids were used as soaps, although, today, most soaps and many cleaning agents are made with synthetic substitutes, thus, they are called "soapless soaps" or "detergents."

There are also quite a few other organic substances in the natural biotic environment with a variety of functional groups; they, too, find diverse applications. The Earth's flora and fauna produce a wide variety of chemical compounds that promote growth, develop, and protect them from environmental threats, such as stressful changes in external conditions (i.e., distress caused by disruptions of the natural balance), dealing with attacks by parasites and/or predators, enabling the transmission of signals (like sexual attraction), etc. This is how they produce phytochemicals such as antioxidants, which are used to capture free radicals, antiviral and antibacterial substances, vitamins, fragrances (pheromones), and more [4].

Cellulose

Cellulose is a linear biopolymer from the family of "polysaccharides" (i.e., complex carbohydrates containing thousands of D-glucose units joined together). It is the most abundant of all the naturally occurring organic compounds, and is present in a wide variety of living species, from bacteria to plants and animals that synthesize cellulose as an extracellular polymer for various biological functions and mainly use it as reinforcement material in plant cell walls. Furthermore, 33% of all vegetable matter on Earth is made of cellulose, while wood and cotton comprise 50% and 90% of the cellulose, respectively.

Cellulose was first described in 1838 by French chemist Anselme Payen (1795–1871) as a resistant fibrous solid residue that remains after the treatment of various plant tissues with acids and ammonia, and after subsequent cellulosic extractions by means of water, alcohol, and ether. Cellulose was used thousands of years ago as a building material and an energy source (in the form of wood and other plants), a source of fibers for clothing and other uses (in the form of cotton and other plant fibers), and when preparing Egyptian papyri. Cellulose has been incorporated into man-made inventions since the Chinese first developed paper in the year 100 A.D. To turn plant material into cellulose for paper, raw plant material is mechanically and chemically broken down into smaller, soluble fibers that can be recast as paper, fibers, or membranes [5].

Likewise, cotton, another important source of cellulose, has been twisted and spun into fibers to form clothing for thousands of years. Today, most of the cellulose used is of man-made cellulosic fibers (MMCF) or regenerated cellulosic fibers manufactured from wood-pulp feedstock, also commercially known as "rayon" or "viscose." Such cellulosic fiber is produced using hydrogen sulfide to break down the naturally occurring biopolymers, lignin and hemicellulose, in the wood, thus releasing pure cellulose. Recently, the application of cellulose was broadened, and novel hybrids and advanced materials were prepared, among them "carbon nanotubes" (i.e., nanoscale tubes made of carbon) and "graphene" (two-dimensional monolayers of carbon atoms) for use in sensors and catalysts. These processes require energy, produce waste products, and leave sewage sludge residue, with some negative environmental effects—though less than those of most synthetic fossil-fuel-based fibers. A 2015 life-cycle risk assessment and environmental health and safety roadmap of cellulose processes identified potential risks from the inhalation of powdered cellulose nanoparticles, mainly in the workplace, that may cause inflammation [6].

Biodiversity

Life is a unique characteristic of the Earth, and biodiversity is the primary characteristic of life. "Biodiversity" is defined as a measure of the stability of the natural environment, and includes the total number of animals, plants, and microorganisms within a given area and their interactions. This index also refers to interspecies' genetic variations, which sets species apart and enables their distinct adaptations to their habitat environments, safeguarding and their continued existence—despite differences between ecosystems, distinct sociocultural populations and their needs, and the necessity to maintain the stability of all those systems. This diversity is necessary for the existence of "ecosystem services" and for the continued existence of life on Earth.

   The basic natural elements of life and biodiversity are the large selection of elements, molecules, substances, and materials used as nutrients and building blocks by all plants and animals, so crucial to their healthy functioning. To review briefly, cells are the basic units of all living creatures, unicellular creatures, and multicellular organisms. All cells have rigid walls surrounding them, built from various polysaccharides (e.g., cellulose and chitin), that protect and support them. Under each cell wall there is a membrane that separates between the external environment and inside of the cell, enabling the selective two-directional passage of resources, inside and out. These membranes consist of a double layer of phospholipids (i.e., lipids with hydrophilic heads, a phosphate group attached to alcohol, and a hydrophobic tail of fatty acids, in which the heads are directed toward the aqueous environments inside the cell from the outside and towards the outside of the cell from the inside). The inner part of the cell consists of "cytoplasm," which contains between 70% and 90% water, in which are found the various organelles of the cell, and where most of the internal processes take place. One of the main organelles in the cell is the "ribosome," responsible for the production of all the cellular proteins that perform the cell's processes. Most cells also have protein filaments that form the cell's skeleton; and their main function is to stabilize the cell structure and enable the movement of all the organelles within the cell and of the entire cell. Many "eukaryotic cells" have a nucleus as well, which holds all the genetic codes (the "genome"), inside, arranged in the form of constructed chromosomes with DNA coils. "Prokaryotic cells," that is, without nuclei, keep their genetic loads in the cytoplasm. Finally, a cluster of cells with intercellular material forms tissues that serve certain functions, like nerve tissue or muscle tissue. Such tissues create dedicated organs with structures suited to the performance of specific actions, such as the creation of an animal's heart or liver, or a plant's leaf or flower—demonstrating the vast amount of physical biodiversity and functional chemodiversity even within the entirety of each individual living organism.

Biosphere

The Earth's biosphere or "ecosphere" contains all our planet's ecosystems and, as such, includes the entire animate environment (all the species in Nature: Flora, Fauna, Fungi, Protista, and Monera), as well as part of the inanimate environment. It is a closed system that includes all our world's minerals, water, gases, and energy by matter. The biosphere also contains a wide variety of locations that differ in their chemical and biological compositions, and environmental conditions, ranging from forests and lakes to savannahs and deserts. One of the main components of the biosphere is its "biomass" (i.e., all the live, biodiverse organic matter in Nature that has not been fossilized). Over the years, various studies have been done to estimate the weight and composition of the biomass in different regions, and similarly in the whole world. The mass of the entire biosphere is often calculated according to the mass of carbon, without accounting for the large quantity of water within the various species. Recently, research estimated the total mass of living things on Earth at ~550 * 10^9 tons of carbon. According to this study, the bulk of the mass of the biosphere is above land (86 wt%), another part is in the depths of the Earth (13 wt%), and the rest in the oceans (1 wt%) [7]. In addition, this mass has also been divided according to the various types of species, as follows: plants (82.5 wt%); bacteria (12.8 wt%); fungi (2.2 wt%); microorganisms without cells (1.3 wt%); protists containing inanimate nucleated organisms that are not plants, fungi, or animals (0.7 wt%); other creatures (0.4 wt%); and the rest for viruses—with the human species making up only 0.01 wt% of the Earth's total biomass.

The natural and human environments serve as homes for a wide variety of living creatures, each one of which has its specific function in the fabric of life. Plants, like algae, serve as the lungs of the world and were the first food producers. The animals, among other things, regulate the quantity and quality of the various species in Nature. The trees, for example, in addition to their important role in the food chain also serve as air filters, temperature and humidity regulators, noise absorbers, and preventers of soil erosion, as well as providing natural habitats for many diverse populations. Another example is the variety of insects that pollinate the plants.

Natural biodiversity also ensures Nature's ability to continue to exist by adaptation to changes. For instance, in the case of vegetarian "nematodes" (i.e., a type of roundworm known to be one of the most harmful plant parasites), the predatory nematodes secrete a toxin that damages the circulatory system of certain insects. This biological function of the predatory nematodes led, in recent years, to their widespread use as biological pest control for insects that damage plants, replacing various toxic chemical insecticides. Scout ants, for instance, mark foraging trails with pheromones, and tunneling ants aerate the soil, and ants break down substances, recycling the decomposed matter back into Nature. Meanwhile, dung beetles and flies decompose manure and rid the soil of animal droppings and food residues, so that garbage will not prevent the soil from regenerating and growing crops. Even the cockroaches, generally considered pests (because they damage plants and transmit diseases), have a role to play like every living creature. Within the food chain, they serve as prey for birds and various rodents, and are even eaten by humans in some places, due to their relatively high density and availability of proteins. Nonetheless, cockroaches play a central role in the nitrogen cycle, a key element in biosystems. Cockroaches usually feed on rotting organic matter (mostly plants), containing many proteins and much nitrogen, and later release that nitrogen with their feces. Then, in conjunction with the activity of certain bacteria, the recycled nitrogen fertilizes the soil for more plant growth. In fact, the bacteria found in cockroach stomachs enable the digestion of various plants that even mammals are unable to digest. However, those cockroaches found in the human environment (e.g., the German and American cockroaches) have gradually adapted to the urban domestic environment, living, and breeding there easily, despite having lost most of their ecological functions.

Endangered Species

According to the data in the *Red Book* of the "International Union for the Conservation of Nature and Natural Resources" (IUCN), which has collected data since 1964 on various threatened species of animals and plants—as of 2021, 40,084 species (of the 2,130,023 described species) are now "endangered species"—that is, ~2% of all the known species (Table 4.2) [8]. However, only 142,577 of those "described" species were, in fact, surveyed in depth, and it was found that ~28% of those are, in fact, also in danger of extinction. That being the case, it may be assumed that the percentage of threatened species, out of the total number of species in the wild, is much greater than 2%. Furthermore, the list of described species (known to exist today) is probably incomplete.

In a later, more detailed survey, it was found that of the 91% of the 6578 described mammalian species actually surveyed, 1333 species were in danger of extinction (20%); of the 100% of the 11,162 avian species that were surveyed, 1445 species were endangered (13%); of the 1.1% of the 1,053,578 insect species surveyed, 2270 species were endangered (0.2%); and of the 15% of the 369,000 flowering-plant species surveyed, 22,477 species were danger of extinction (6.1%). Moreover, according to the IUCN, in the absence of significant action, the forecast is that 99.9% of the critically endangered species and 67% of the endangered species will be extinct within 100 years.

Many other studies have been carried out in recent years to assess the biodiversity crisis, and they all point to the great depletion of many species. One report claimed that 40.7% of the "amphibians" (i.e., cold-blooded vertebrates, like frogs) and 25.4% of the mammals are threatened by extinction, as are 13.6% of the birds [9]. Yet another, more recent, report found that at least 1829 (21.1%) of the 10,196 species of "reptiles" (i.e., scaly-skinned vertebrates, like snakes) are also threatened [10]. In fact, it was discovered that in many species, even some that had never been on the "endangered species" list—a very high proportion of their populations had, indeed, become extinct. For example, a recent study estimated that 27,600 vertebrates are suffering significant population loss [11]. Moreover, of the 177 mammalian species examined, it was found that more than 30% of them had lost their usual, natural habitats, and more than 40% had experienced reductions of more than 80% of their populations. In total, it was estimated that 50% of the number of vertebrates

Table 4.2 The number of threatened species out of the total number of described species by main groups of organisms

	Described species	Surveyed species	Threatened species
Vertebrates	73,883	57,155	10,437
Invertebrates	1,491,386	26,514	6042
Plants	423,373	58,343	23,335
Fungi and primitive organisms	141,381	565	270
Summary	2,130,023	142,577	40,084

that had lived on Earth had disappeared. Finally, note the greater significance of species extinction. Along with species extinction, certain genes also disappear from the pool of genetic diversity, thus increasing the risk that they will never be regenerated, and markedly increasing the danger of further species extinction.

Mass Extinctions

Starting with the appearance of the first microorganisms in Nature, ~3.5 billion years ago, a continuous evolutionary process began, during which a great variety of species multiplied, changed, formed, and dispersed to different ecosystems on the surface of the Earth, initially in the water and later also on land. Meanwhile, alongside the prosperity of existing species and the appearance of new ones, and as part of the ongoing evolutionary processes, it is likely that these and other species became extinct. Such processes span thousands, hundreds of thousands, and even millions of years. Both processes, the creation of new species and the extinction of species are integral parts of the process of natural selection. Thus, the total population of species in Nature, with reference to the absolute number of species and the amount of each species, is not fixed, and is constantly changing.

The "background rate" of species extinction over time, as calculated from fossils, stands at 0.01–1 extinctions per million species per year [12], which is equivalent to the extinction of 1–3 families having several species. Every million years, when certain taxonomic families become "extinct," it means that all the individuals in those particular families no longer exist. For instance, according to estimates, the background rate of extinction of birds is one species per 400 years, while that of mammals is one species per 700 years [13]. At the same time, new species are being created at a rate of 0.02–0.5 species per million species per year [12], thus maintaining near equilibrium in the number of species in Nature. Nevertheless, when the extinction rate of existing species is very high, relative to the creation rate of new species, the natural balance is disrupted, and a catastrophic event, known as an "extinction event" occurs.

Extinctions of species, including "mass extinctions," have occurred throughout Earth's history, some set in motion directly (as by geological or geomorphological processes) and others indirectly (by genetic or environmental causes), or by a combination thereof. In general, it may be said that any species that cannot adapt to its environmental conditions, to survive and reproduce or, alternatively to find and move to a new, viable habitat, will become extinct as part of Darwin's "survival of the fittest theory" [14]. Extinction, resulting from both natural and man-made processes, may be sudden or occur gradually, and happen locally or on a global scale.

Since the prehistoric appearance of the first species, there have been quite a few extinction events, but when a certain extinction event is both widespread and prolonged, it is termed a "mass extinction event." Today, it is generally accepted that during the history of life on planet Earth, before the appearance of the human species, there were five mass extinctions, although there are those who claim that there have been as many as 20 mass extinctions. The five attested mass extinctions occurred: (1) ~450 million years ago, during the transition from the Ordovician to the Silurian period, when 57% of the "genera" (pl. of "genus") and 86% of the species became extinct; (2) ~360 million years ago, during the transition from the Devonian to the Carboniferous period, when 35% of the genera became extinct and 75% of the species; (3) ~250 million years ago, during the transition from the Permian to the Triassic period, when 56% of the genera and 96% of the species became extinct; (4) ~200 million years ago, during the transition from the Triassic to the Jurassic period, when 47% of the genera and 80% of the species became extinct; and (5) ~65 million years ago, during the transition from the Cretaceous to the Paleogene period, when 40% of the remaining genera and 76% of the remaining species became extinct.

As mentioned above, the causes of such mass extinctions are diverse. Some hypothesized causes are impacts by celestial bodies striking the Earth (such as asteroids or meteorites), outbreaks of ionizing electromagnetic gamma radiation originating from other planets, volcanic eruptions, earthquakes due to the movements of the tectonic plates (clashing or separating), or climatic changes that alter temperatures, atmospheric and marine oxygen concentrations, and so forth.

The most extensive and severe extinction event, the Permian-Triassic extinction event (P-Tr), also known as the "Great Dying," lasted several thousand years, and involved the extinction of ~80% of all the marine species and ~70% of the terrestrial ones, both flora and fauna [15]. There are several predominant theories about the causes of the P-Tr event, ranging from a meteorite hit to climate changes caused by the formation of the supercontinent Pangea ~335 million years ago, which led to changes in the wind regime on land and in the ocean currents. However, today, the leading hypothesis is that an unprecedented volcanic eruption had occurred in the region now known as central Siberia, which had lasted for about two million years, causing that "mass extinction event." It is estimated that during this eruption more than 200 billion liters of molten lava were ejected over an area of more than two million square kilometers, in the area known as the "Siberian Traps." That eruption also caused the emission of many gases into the air, including GHGs (such as vapor water and carbon dioxide), which caused significant warming of the atmosphere [16], as well as toxic gases (like sulfur dioxide and hydrogen sulfide). It is estimated that shortly after the eruption, the Earth experienced significant climatic changes, and the surface of the oceans warmed by ~11 °C, which, in turn, led to the extinction of marine species. At the same time, the huge volume of sulfur dioxide and carbon dioxide released into the atmosphere led to a lack of oxygen in the air and to the formation of "acid rain" (sulfuric acid), produced by a reaction between water vapor and sulfur dioxide.

However, there are researchers who claim that the estimated rise in ocean water temperature alone does not explain the extent of that extinction event, rather there must be another simultaneous, indirect phenomenon that joins it, termed "anoxia" (i.e., the depletion of the oxygen concentration in the water, which is essential for most marine creatures) this due to the change in the water's solubility caused by the raised water temperature.

The fifth mass extinction, the Cretaceous-Paleogene (K-Pg) or Cretaceous-Tertiary (K-T) extinction, took place ~65 million years ago and resulted in the loss of more than 75% of the flora and fauna on Earth [17]. This extinction is also well known for the extinction of the dinosaurs, who ruled the Earth for almost 150 million years, although they were only a small part of the biodiversity lost at that time. The K-Pg was a more selective extinction, in which the depletion of species in the tropics was much more severe than in other regions and there was also significant variation in the rate of extinction of the different families and within them. Although many hypotheses have been proposed over the years in the attempts to explain the fifth extinction, ranging from the emission of toxic volcanic gases and the scattering of dust to climatic changes that led to food shortages, there is now a consensus that the cause of that extinction was a fireball, meteorite, or comet that hit the Earth and caused the dispersion of a huge amount of hazardous *dust waste* in the atmosphere. This dust enveloped the Earth in darkness for several months or more, resulting in the death of green plants due to the lack of photosynthesis, which also disrupted various food chains [18].

The different extinctions led to a significant decrease in species' variety in Nature, while allowing previously marginal species (that had been stymied by the previously dominant species in their area) to explore and diversify. Thus, for example, the disappearance of the dinosaurs was the beginning of the development of mammals. However, millions of years passed until the number of species in the various regions returned to the numbers that had been there before the mass extinctions [19].

The Sixth Extinction

Although it is customary to state that only ~2% of the species that have ever lived are still alive today, the absolute number of species that existed in Nature with the appearance of Humanity was probably greater than ever [20]. However, as humans began to prosper and conquer new areas, it caused direct and indirect damage to more and more habitats and ecosystems, resulting in the depletion of ever-growing numbers of species. Due to the accelerated growth of the global human population over the past centuries and our higher standard of living during the last few decades, the irreversible damage we have done to all the Earth's natural species is now on the order of mass extinction. Unlike all the mass extinctions listed above, which occurred before the appearance of Humanity and were caused by natural processes, the current mass extinction, known as the "Sixth Extinction," in which we find ourselves, is not only the first mass extinction in the "human era," but also the first man-made extinction [21]. Moreover, relative to all the previous five mass extinctions, this one is happening very rapidly. It is estimated that the current species mass extinction rate is a hundred to a thousand times higher than the background rate of extinction [22, 23], and, some claim, even 10,000 times higher [24]. Furthermore, it encompasses not only many species but also involves diverse ecosystems in large areas on Earth.

A report of the World Wildlife Fund for Nature (WWF) from 2020 reported an ~68% decrease in the number of mammals, birds, fish, reptiles, and amphibians between 1970 and 2016, based on a survey of ~21,000 populations [25]. It was also reported that, in 2019, the tropical rainforests were disappearing at the rate of the area of a standard American football field (i.e., 57,600 ft = 17,556 m) every 6 s! However, a study published at the beginning of 2022 claims that the depletion rate of various species populations is much higher, since the WWF had only surveyed animals in protected areas and Nature reserves; if the depletion there is significant, then the depletion in the wild natural environment is surely greater [26]. The conclusion that emerges from all the relevant surveys and studies is that this ongoing Sixth Extinction is the result of the combined damage from the destruction and loss of natural habitats in favor of agriculture expansion; increased urbanization; human overexploitation of natural resources; chemical pollution of ecosystems by modern industrial processes; climate change caused by man-made GHG emissions; and invasions by invasive species. Briefly stated—the current Sixth Extinction is far more severe than anything thus far reported in the existing assessments.

Wolves

Each distinct type of species plays its part in the ecosystem in which it exists; therefore, even the loss of any one species or another will cause a "ripple effect" on other species and/or on the entire ecosystem, even if it is not a key species in that ecosystem. As mentioned, although the loss of certain species over time does occur naturally due to local or global changes—since the appearance of Humankind, and with the growth of the human population, and the expansions of agriculture and urbanization—it is Humanity that is causing most of the damage to the other species. This is true, whether it is being done directly (e.g., by damage to habitats, hunting, and forest fires) or indirectly (by the transfer of invasive species and negative changes caused to the natural environment, and to the Earth's climate).

One of the most well-known and oft cited examples of the influence of one specific species on the fabric of Nature itself is that of the influence of the wolves in Yellowstone National Park in Wyoming on the biotic and abiotic environments there. The wolves that had lived in the area for a long time were hunted to the point of extinction, and were no longer seen in the area after 1930. As a result, the deer in the park, with no natural enemies around, flourished, so much so that they consumed all the prairie vegetation and then killed the trees, especially those on the banks of the narrow river. This damaged the durability of the riverbanks, enabling the river to widen, reducing the number of insect-eating birds in the area, letting the mosquito and other insect populations multiply. In other words, the Yellowstone ecosystem was collapsing. When wolves were returned to the park in 1995, with hopes of restoring balance to this dying ecosystem, they preyed on the deer and, once again, the vegetation flourished and, with it, the birds, beavers, fish, and other animals returned to the banks of the river and in the river itself. With the riverbanks restabilized, the river followed a fixed route. Both the local geography and climate were restored. This process of removing or adding a predator at the top of the food chain is called a "trophic cascade" and has cascading, far-reaching effects on the entire natural environment.

Invasive Species

Today, one of the main factors affecting the diversity of species in Nature and changing the natural local environmental balance beyond recognition is the take-over and establishment of invasive species—those that reach areas located outside of their natural distribution. These species often break the balance of the local natural systems, harm the local species, directly or indirectly, and cause great damage to both the ecosystems and to humans.

While, due to local and global ecological changes, different species move from place to place naturally, the socioeconomic activities of humans accelerate the migration of many species, for example, via the transportation of goods or the invasion of habitats, obliging the lower local species to find new living spaces. Such invasive species (like Humankind) often manage to establish themselves, reproduce, and thrive in their new locations, even further enlarging their areas of distribution. Moreover, in the absence of their natural enemies, these invasive species—living in local ecosystems that cannot naturally balance the invaders' population by curbing its proliferation—may reproduce and spread at a higher rate and aggressively cause heavy damages. Various studies in this era show that about a third of the animal species and about a quarter of the plant species became extinct due to invasive species that destroyed their habitats and ruined their supporting ecosystems [27]. Invasive species also cause damage to human economy, health, and the quality of human life.

Zoonotic Diseases

Humanity's invasions and damage to natural environments and species' variety, along with the massive human consumption of many species as food, as well as the production of certain man-made products, also led to close contact between humans and other natural species—and, as such, diseases were transmitted from humans to the different species in Nature and vice versa. From the dawn of Humanity to our present, diseases and epidemics, such as typhus, cholera, Ebola, AIDS, SARS, and COVID, have struck humans. These diseases, transmitted to humans from animals, either directly or indirectly, are termed "zoonotic diseases," may cause death and have far-reaching socioeconomic effects [28]. The close contact between humans and animals found at home, on farms, in industry, and in Nature, enables many disease-causing bacteria, viruses, parasites, and fungi to pass from the animals to humans, causing illnesses that can be mild but may also be life-threatening.

There are several human factors involved in human infection by zoonotic diseases. The primary factor is Humanity's constant increase in the consumption of animal protein; zoonotic diseases may be transmitted by animals raised to be food and/or by the food itself. For example, more specifically, the processing of meat and the entire food chain of animal protein, "from the farm to the plate," can lead to the proliferation and outbreak of many diseases. Another potential source of contagion is raising animals on farms under overcrowded conditions, often in a dirty environment, which leads to viral outbreaks. Furthermore, the relatively low genetic diversity of domesticated animals is fertile ground for the establishment and outbreak of viruses that cause infectious diseases. Even trade in hunted wild or exotic animals is a major factor in the transition of zoonotic diseases from remote habitats to the human environment, endangering human health. The fact that today's world is "global," and there is constant movement between continents, regions, and countries, leads to the rapid transition of diseases from one place to another. Finally, global warming dramatically affects both viruses and their host animals, with generally warmer and wetter areas found to be suitable for many viruses. At the same time, global warming also weakens the human immune system, making it more vulnerable.

Zoonotic diseases can pass from animals to humans in several ways. There are diseases transmitted by direct contact with saliva, blood, or animal secretions (e.g., rabies often transmitted by a dog bite). Others are transmitted by indirect contact, as when a bacterium or virus passes from animals into the environment; or a disease from a wild animal infects another animal, usually a domesticated one, from which it is further transmitted to humans (e.g., as in cases of bird or swine flu). The third common way is through a host, known as a "vector" (e.g., a mosquito, flea, etc.). In the case of the parasitic protozoan, *Leishmania*, the single-cell organism that causes Leishmaniasis—it is transmitted from Hyraxes (mammalian "rock rabbits") to humans via the sand fly (the vector), just as malaria is transmitted by certain mosquitoes. Moreover, it is possible to contract the disease through the food or drink we consume, from Salmonella that passes through food, to leptospirosis, which originates from the *Leptospira* bacteria, often secreted by mice and rats, which sometimes reach water sources used by humans. It is also important to note that diseases passed from animals to humans can not only be highly contagious and deadly but, over time, various mutations also develop in humans, resulting from changes in the genetic compositions of those viruses, making infections even more dangerous and/or contagious. "Pathogenic diseases" (i.e., diseases caused by infectious agents that invade a body) can also pass from humans to animals by the same mechanisms described above.

One Health

The concept of "health" is very complex and, in recent years, alongside the diagnosis and treatment of diseases, other aspects of health awareness and prophylactic medicine are being discussed and developed, such as genetic screening, the promotion of healthy lifestyles, improving the quality of life and longevity, and so on. However, quite recently, a more advanced concept has been proposed, which discusses health in a much broader, holistic manner, based on a combination of human, animal, and environmental health, all falling under the WHO "One Health" model [29].

The relationship between the health of the environment and the health of animals and human beings is both complicated and reciprocal. Both humans and animals (domesticated or living in the wild) share quite a few resources and are affected by the changes in the natural environment. Meanwhile, climate change and human activity, especially urbanization, agriculture, and industrialization, have been and are still causing significant damage to the natural environment, ecosystems, and habitats—affecting the health of animals and humans, alike. For instance, all the above increase the potential for outbreaks of epidemics in humans. The "One Health" concept seeks to combine experts from all the relevant fields—medicine, veterinary medicine, ecology, and others—and to merge the information, knowledge, and expertise, thus integrating and focusing both the professional experts and the data on the issue of how to best manage environmental and personal health optimally. As such, this process requires shifting from the "anthropocentric approach" (that puts humans and their needs in the center), to the "biocentric approach" (which considers humans to be equal members in Nature), emphasizing the equal value of the lives of all living beings, even to the point of "ecocentrism" (that claims the right of existence for all the components of Nature, all its biotic and abiotic members).

References

1. Dose K (1998) Prebiotic evolution and the origin of life: chemical and biochemical aspects. In: Jeanteur J, Kuchino Y, Müller WEG, Paine PL (eds) Progress in molecular and subcellular biology. Springer, Heidelberg, pp 97–112
2. Nazarevich VJ (2015) Earth Common J 5:61–72
3. Roberts JD, Caserio MC (1977) Basic principles of organic chemistry. WA Benjamin Inc, New York
4. Caron S (2020) Practical synthetic organic chemistry: reactions, principles, and techniques. Wiley, Hoboken
5. Garland NT, McLamore ES, Gomes C, Marrow EA, Daniele MA, Walper S et al (2017) Hybrid polymer composite materials. Woodhead Publishing, Cambridge, pp 289–320
6. Shatkin JA, Kim B (2015) Environ Sci Nano 2:477–499
7. Bar-On YM, Phillips R, Milo R (2018) Proc Natl Acad Sci U S A 115:6506–6511
8. https://www.iucnredlist.org/resources/summary-statistics. Accessed Jul 2025
9. Boillat S, Speranza CI (2019) PBES secretariat, Bonn
10. Cox N, Young BE, Bowles P, Fernandez M, Marin J, Rapacciuolo G et al (2022) Nature 605:285–290
11. Ceballos G, Ehrlich PR, Raven PH (2020) Proc Natl Acad Sci U S A 117:13596–13602
12. De Vos JM, Joppa LN, Gittleman JL, Stephens PR, Pimm SL (2015) Conserv Biol 29:452–462
13. Shivanna KR (2020) Resonance 25:93–109

14. Darwin C (1964) On the origin of species: a facsimile of the.1st edn. Harvard University Press, London
15. Pope KO, Baines KH, Ocampo AC, Ivanov BA (1994) Earth Planet Sci Lett 128:719–725
16. https://www.britannica.com/science/K-T-extinction. Accessed Jul 2025
17. Ceballos G, Ehrlich PR, Dirzo R (2017) Proc Natl Acad Sci U S A 114:E6089–E6096
18. Jablonski D (2002) Proc Natl Acad Sci U S A 99:8139–8144
19. Raup DM, Sepkoski JJ (1982) Science 215:1501–1503
20. Chapagain AK et al (2006) Ecol Econ 60:186–203
21. Pimm SL et al (1995) Science 269:347–350
22. Pimm SL, Jenkins CN, Abell R, Brooks TM, Gittleman JL, Joppa LN et al (2014) Science 344:1246752
23. Wang ZM, Pierson RN, Heymsfield SB (1992) Am J Clin Nutr 56:19–28
24. Almond REA, Grooten M, Peterson T (2020) Living planet report 2020-: bending the curve of biodiversity loss. World Wildlife Fund, Gland
25. Murali G, de Oliveira Caetano GH, Barki G, Meiri S, Roll U (2022) Nature 601:E20–E22
26. Díaz SM, Settele J, Brondízio E, Ngo H, Guèze M, Agard J et al (2019) The global assessment report on biodiversity and ecosystem services: summary for policy makers. IPBES secretariat
27. Blackburn TM, Bellard C, Ricciardi A (2019) Front Ecol Environ 17:203–207
28. Rahman MDT et al (2020) Microorganisms 8:1405
29. World Health Organization. One Health. https://www.who.int/health-topics/one-health#tab=tab_1. Accessed Jul 2025

Chapter 5
The Human Environment:
The "Anthroposphere"

Abstract Population growth and the rise in quality of life have led the human society to establish a new environment—the Anthroposphere—where most human social and economic activity takes place. This human-made environment was shaped by major revolutions, particularly the "Agricultural Revolution" and the "Scientific-Industrial Revolution." Meanwhile, activity within the Anthroposphere has influenced and even reshaped the natural environment—living (animate) and inanimate. Moreover, the development of economic and sociocultural models and processes, prioritizing economic growth beyond the limits of natural systems (especially consumer culture, which has become the dominant culture of our time) has led to the distancing, even dissociation, of Humanity from Nature, resulting in damage to natural resources and ecosystems. In this context, the profound and widespread negative changes in natural resources, the quantities and types of various chemicals, both lost and found in Nature, along with the negative changes in biodiversity, have recently prompted calls for the determination of a new geological epoch—the Anthropocene—reflecting the fact that the human species is now (mis) shaping our Planet.

Keywords Anthroposphere · Population growth · Consumer culture · Anthropocene

Like all the other species, the Humankind developed through a very long evolutionary process that began ~30 million years ago with the appearance of the "great apes," followed by the appearance of the upright "Australopithecus," the first "hominid" species ~5 million years ago. Then, some 2–2.5 million years ago, those upright hominids, our forebears, would ultimately evolve into the first human beings (i.e., "*homo sapiens*"). Initially, there had been other humanoid species, "*homo habilis*" and "*homo erectus*," which had both evolved in Africa, but became extinct. The first humanoid species thought to have appeared in Africa ~200,000 years ago or even earlier was "*homo sapiens*" or the "wise human," credited with being the first to use fire and stone tools. Later, "*homo sapiens*" eventually left Africa and migrated to the Middle East and Europe. Note that ~130,000 years ago,

A. Wolfson, *Chemodiversity and the Ecological Crisis*, SpringerBriefs in Molecular Science, https://doi.org/10.1007/978-3-032-07363-1_5

"Neanderthals" (also humanoids) had evolved outside Africa but had also become extinct. Finally, ~40,000 years ago, a sub-species of "*homo sapiens*"—the "*homo sapiens sapiens*" or "modern wise man"—from which we are all descended, evolved, and became the only humanoid species that survived.

Since the appearance of the first human being on Earth, hundreds of thousands of years ago, *homo sapiens* has multiplied, developed, and prospered. As one of the species in Nature's biodiversity, in the animal kingdom and in the Mammalia class, which is also among the top predators, the human species has unique characteristics and many roles. But during the millions of years that have passed since the appearance of the different species in Nature, and from the moment when the human species appeared, and even more so from the moment when the subspecies of the "modern wise man" (*homo sapiens sapiens*) developed, it became a dominating, invasive species, spreading on Earth rapidly, in an unprecedented manner, conquering the land, the sea, and even reaching into outer space. Over the course of history, Humanity has experienced three key "revolutions": (1) the development of "proto-language" either by *homo habilis* ~2.5 million years ago, or by *homo erectus* ~1.8 million years ago, and proper language by *homo sapiens* ~200,000 years ago; (2) the transition from a nomadic society of hunter-gatherers to a sedentary agricultural society that clings to the land in permanent settlements ~12,000 years ago; and then, (3) becoming an industrialized, urban society since the late-eighteenth century, that distanced and disconnected humans from their part in Nature; eventually transforming *homo sapiens sapiens* from a species that not only dominates the inanimate and animate environments, but also took overwhelming control of Nature itself, thus altering (and disrupting) natural processes.

Initially, the inclusion of Humankind in the "Hierarchy of Nature" had been clear; there was an implicit understanding and belief in a certain order in the world. This order is reflected in the cycles of natural phenomena, such as day and night, the changing seasons, reproductive cycles and mortality, and cycles that sustain the biotic and abiotic systems. These natural cycles, alongside the ecosystem services that Nature provides, maintain the existence of all the species, including humans, by supplying vital resources (e.g., food, water, and air); regulation of environmental conditions (like humidity and temperature); and supporting services (such as, plant pollination, decomposition of organic matter, etc.)—ultimately determining the nature of human life. In fact, the basic needs, the way of life, and the survival of our ancestors, who were hunter-gatherers, depended on the food, shelter, and tools that Nature provided them directly, when they, in return, provided Nature with various services, for example, thinning the flora and fauna and transporting seeds from place to place.

The first period in human history began ~2.6 million years ago and is known as "prehistory" (i.e., the period before "history" that began with the invention of writing ~4000 years ago). The outset of prehistory is usually marked by the start of the regular human production of tools and their use for the processing of natural materials for the benefit of Humanity. This prehistoric period is divided into three sub-periods, represented by the primary material used by Humankind to fashion those tools at that time: the Stone Age, the Bronze Age, and the Iron Age. The beginnings and ends of these sub-periods were different from place to place and, in some places, they even occurred in a different order. In fact, there were places, such as the Middle East, where another (fourth) period was distinguished—a transition period between the Stone Age and the Bronze Age, in which both materials were used simultaneously, initially termed "the Copper Age"; however, it is currently known as "the Chalcolithic period" (from the Greek, a compound term consisting of the words copper—*chalcus* and stone—*lithos*). Moreover, today it is clear that in some periods, tools from earlier periods were still in use, so that these divisions are not absolute.

With the development of the human species, revolutionary man-made processes were added alongside the natural evolutionary processes. These "revolutions," driven by significant and extensive technological changes, time after time, altered the way humans perceived themselves and their environments, leading to far-reaching socioeconomic and environmental changes [1]. But more than anything, these human revolutions repeatedly redefined Humanity's perception of Nature—shifting it from an animistic viewpoint (which ascribes a soul to Nature's living and inanimate components, seeing each one as subordinate to nature as a whole) to an objectifying perspective that considers Nature to be lifeless (without a soul), enabling human detachment and distance and, thus, creating a sense of ownership with the right to appropriate. This perceptual change also redefined Humanity's use of the Earth's material and energy resources regarding food, heating, clothing, housing, movement, and communication and, accordingly, also regarding the amount of material in Nature, its quality and type: natural or synthetic, raw or processed, organic or inorganic, and so forth. Humankind repeatedly changed the natural environment, animate and inanimate, while defining a new environment—"the human environment."

The dynamic relationship between humans and Nature changed significantly during the Neolithic period, the "agricultural revolution," ~12,000 years ago [2]. This revolution, characterized by the cultivation of plants and the domestication of animals, changed the way of life of the hunter-gatherers, who had been nomads, transforming them into farmers living in permanent settlements. Additionally, due to the increased growth of certain species of plants and animals and the natural, material resources involved, a massive amount of biomass was being produced in a place other than its natural place. In short, this agricultural revolution also changed the abiotic environment. Humans expropriated, fenced, and cultivated natural areas, diverted streams, expanded pastures, and more. Meanwhile, beside the conversion of open, natural areas in favor of agriculture and the increased, biased use of natural resources for the needs of human society, human activity had begun the pollution of the soil, water sources, and air—whether with the intention of burning vegetation to clear land for agriculture, or by burning biomass for heating and cooking, or due to the discharge of sewage. All these human actions caused damage to biodiversity, natural cycles, and ecosystems, as well as to Nature's ability to renew its own resources, even eventually disrupting various evolutionary processes.

Indeed, the agricultural revolution led to trade and the development of a "leisure culture" (i.e., one that makes pleasure-seeking, hedonistic demands), which constantly increased its use of natural resources, not only to meet basic needs (for food, clothing, and shelter), but also to make more tools, more clothing, ornaments, and jewelry. Moreover, the accompanying transition to permanent settlements, with the aim of holding the agricultural land and storing the crops, and the urbanization processes that followed, made greater and greater demands on the use of natural resources for construction, the establishment of infrastructures, and the production of energy, again leading to even more production of polluting waste products, effluents, and air emissions [3]. All these processes defined the human environment for the first time not only as an environment expropriated from Nature and in which there is a high density of humans and increased human activity, but especially as one in which Humanity controls the natural resources. Accordingly, the ratio between the size of the human population and the natural materials being processed and consumed by Humankind (agricultural products, building materials, and accessible raw natural materials) keeps increasing.

These man-made processes continued to become more sophisticated with the advent of the "scientific-industrial revolution," as the amounts and types of materials also changed. New definitions arose delineating the natural and human environments. The scientific revolution during the sixteenth and seventeenth centuries (at the start of the modern era) [4] promoted knowledge and scientific research, and like them their exposure to all, and leading to the determination of distinct scientific disciplines, such as mathematics, chemistry, and physics. This open-minded period produced a wide range of discoveries in different fields, changing the perception of science and its usefulness. This revolution laid the scientific basis for the "industrial revolution," which began at the end of the eighteenth century, and would produce many innovative and ground-breaking technologies, the first of which was the invention of the steam engine [5].

During the industrial revolution, the development of man-made technologies was accelerated, again significantly changing human life, and resulting in massive developments of socioeconomic, and environmental processes, already initiated during the agricultural revolution. For instance, the mass production of food by means of mechanized processes led to a surplus of goods and the expansion of trade. Furthermore, "urbanization" (the construction of buildings and infrastructures) became accelerated, as did the development of technological tools (e.g., machinery for producing textiles or food, new means of transportation, and computers). The industrial revolution also brought the production of artificial or synthetic molecules and materials (such as fertilizers, medicines, steel, and plastic), including many materials foreign to the natural environment, that do not naturally degrade (i.e., that are not chemically broken down by microorganisms). By the late-twentieth century, genetic engineering had also emerged, enabling the external and artificial modification of genes in a certain species, or the transmission of genes from one species to another, to improve the properties of the species by making them more resistant to environmental threats and enhancing their performances (e.g., to produce a greater yield or higher quality products).

The exponential increase in human knowledge gained during the industrial revolution led to the appearance of new technologies that spawned additional sub-revolutions. The "First Industrial Revolution" (in the late-eighteenth and early-nineteenth centuries) was when coal and steam were used in mechanization meant to increase production and promote more employment in agriculture and, especially, to bring more manpower into industry. The "Second Industrial Revolution" or "technological revolution" (in the late-nineteenth and early-twentieth centuries) expanded the production of electricity, advanced the transportation and communication sectors, and discovered new uses for oil and steel, increasing their production accordingly. The "Third Industrial Revolution" or the "Information Age" (from the mid- to late-twentieth century), focused on the uses of advanced electronic, digitization, and information technologies. These made it possible to increase production capacity, mainly via automation, characterized by two main stages—"the digital revolution," primarily shifting electronic and mechanical devices from analog to digital [6] and "the information and communication revolution," introducing novel technologies for information storage and security. Currently, we are amid the "Fourth Industrial Revolution" (recently begun in the early-twenty-first century) that is connecting the physical and digital worlds by means of various information technologies, such as: "Big Data," able to analyze and optimize large amounts of data and make logical predictions; the "Internet of Things" (IoT), which enables different devices to connect with each other and exchange data [7]; "artificial intelligence" (AI), which simulates human thinking ability via technological means; "quantum computing," able to perform extensive, rapid data processing; and actual digital-human interfaces, like: 3D printers, virtual reality, and augmented reality.

There is even a budding, foreseeable "Fifth Industrial Revolution" that seeks to take all this knowledge and these advanced technologies a step further by blurring the boundaries between the physical, the biological, and the digital. It deals with diverse fields, ranging from the creation of "cyborgs" (entities containing biological and artificial components) to the direct linkage of human brains to various machines [8]. All these new technologies and their products require the use of materials and energies, and yield different emissions, adding more stress and new challenges to the natural environment.

The "Anthroposphere"

The "agricultural revolution," with human settlements that supplied their inhabitants with a variety of services ranging from housing, the provision and storage of goods, protection, and trading—these sowed the seeds of the "human environment," the "anthroposphere" (i.e., an artifact generated by the interactions between human social systems and ecosystems). Nevertheless, several thousand more years were needed to develop the required advanced technologies and services (e.g., a potable water supply, sewage and waste removal systems, governance, cultural and medical services, etc.) that characterize modern cities [6]. The populations of ancient cities varied from several thousands, at the dawn of cities, to tens and even hundreds of thousands, or more, at their peaks. The cities contained a combination of infrastructures that satisfied their citizens' basic needs and services (e.g., food, shelter, and security), ranging from employment and commerce to health and entertainment, and a community that shared in the social life and culture. It has always been these very needs and services that shape the man-made urban environment—a human environment mostly constructed of synthetic materials, effectively erecting barriers between people and Nature. By physically distancing people from Nature, this synthetic barrier eventually caused a sense of estrangement between Humanity and Nature. Eventually, cities grew tremendously, almost uncontrollably, as urban architectural design was revised, and factories, warehouses, and offices were added alongside houses and municipal buildings. This aggressive, dominant human environment altered the city landscape, as well as the lifestyles of the urban residents, and even the way that cities were being run. On the national level, the bulk of the economic, cultural, and governmental systems and facilities became concentrated in the major cities. Novel technologies and sociocultural changes were notably altering and reshaping the anthroposphere, further distinguishing the human environment from the natural environment, while constantly altering the relationship between Humanity and Nature.

In comparison to the biosphere, the anthroposphere is characterized by less natural material and more synthetic material, and less biological or organic material and more inorganic material. Nonetheless, despite the intensive use of material resources, the human environment is also an environment rich in energy applications: for producing electricity, for driving vehicles, for industrial uses, for operating computers, and for powering information and communication technologies.

Finally, over the past decade, humans have also developed "virtual environments" [9], using different virtual technologies (i.e., networked applications that create imaginary or realistic worlds that allow users to interact with both the computing environment and the work of other users). In contrast to the physical world, where physical matter is seen, touched, and ingested, the virtual world is a symbolic one, in which computer programs provide functionality and massive amounts of data can be transmitted. Moreover, in the virtual world, time and space, and socioeconomic concepts gain new meanings, since they are not subject to real-world physical laws, nor to existing ethical norms.

A virtual environment requires different types of materials and energies, such as high amounts of metal and plastic, rather than of organic compounds. Moreover, computers and robots do not require sustenance and their cores consist of inorganic materials; as their use in daily life grows and human society becomes more dependent on these advanced machines, so the dependence on the inorganic matter in Nature will increase. Those hi-tech machines will be able to function in environments in which humans cannot exist, such as in the deep sea or outer space, and function under extreme conditions, since they are not subjected to the environmental threats that face Humankind (e.g., air pollution and shortage of potable water). Finally, add "the assumption of the technological singularity," which states that systems of artificial intelligence will be capable of learning, forming conclusions, and acting accordingly autonomously, and much more rapidly than the natural changes occurring in the human mind [10]. If so, "technological evolution" will become much faster than biological evolution [11], which could well lead to another leap forward for human society, despite putting the continued existence of Humankind in great danger. Such a scenario does not seem far from today's reality, though no one knows how to predict if and when such a situation will materialize, and especially what will happen when we reach that point. Will Humanity become extinct or manage to adapt to those changes? The crucial meaning of such a scenario is that "evolutionary" and "revolutionary" processes will have become intertwined, integrated together. In other words, just as the conjoined natural evolutionary processes (biological, chemical, and physical) led, over time, to progressive development and affected the cognitive, sociocultural processes of humans, so have human revolutionary processes led to sociocultural changes impacting the natural processes, until it will be impossible to distinguish between and separate the two. Ultimately, when coupled with the accumulated damage that was done to the natural environment and the geological and biological changes caused by human activity, a new evolutionary process may occur, in which the inorganic matter in Nature replaces the organic matter as the basis of "life" on Earth.

It's Crowded Here

In 1804 A.D., the world population numbered one billion people. Since then, the Earth's population has steadily grown from two to eight billion people at intervals of 123, 33, 14, 13, 12, 12, and 11 years, respectively, and is expected to reach nine, ten, and eleven billion people by 2037, 2055, and 2088, respectively [12]. Despite the enormous rise in the global population, and the fact that over the past hundred years the global population has quadrupled, the numbers show that the rate of population growth peaked (2.1%) in 1968, and now the rate of population growth is decreasing (currently at only 1% per year). Meanwhile, the global fertility rate has dropped from 3.2 births per woman in 1990 to 2.5 in 2019 and is even expected to further decrease to 2.2 by 2050 [12]. A growing number of countries are experiencing shrinking population sizes, since the Earth's human population is aging. Scientific forecasts indicate that, by the year 2050, more than half of the world's population will be concentrated in nine countries, mostly located in Asia and Africa; the population of sub-Saharan Africa is expected to double by 2050. In line with this forecast, the Earth's population growth rate will be zero in 2100!

In 1798, Thomas Robert Malthus (1766–1834), an English demographer and economist, published an essay entitled: "An Essay on the Principle of Population" [13], in which he wrote that the irrepressible urge of humans to multiply will eventually lead to overpopulation of the Earth, to the depletion of all its resources, and to mass death by starvation. Although Malthus' prediction did not come true, and for many years it was seen as a prophecy of wrath and nothing more, years later, when the population increased and the negative impact of the human population on the natural environment, especially on the depletion of natural species diversity, became more and more acute. Questions regarding the maximum number of people that the Earth could sustain began to come up repeatedly.

The term "carrying capacity" represents the number of individuals in a certain species that can exist, over time, in an environment with given resources. It refers to the limited resources within a given area and Nature's ability to continue renewing those resources. Regarding human beings, "carrying capacity" refers to the number of people who can exist in certain area with given resources, without harming the natural environment and its ability to renew its resources, as well as the ability of current and future generations to sustain their lives. Unlike all the other species in Nature, in the case of humans, the carrying capacity is affected not only by the number of people but also by their standard of living. Also, although it is possible to change or improve the carrying capacity of a certain area by technological means, there is no doubt that human activity and the human impact on natural resources and biodiversity are not only local, no longer limited to residential location.

In 1968, Prof. Paul Ehrlich (1932-), a biologist, who for many years dealt with issues of population growth, wrote the book: *The Population Bomb* [14], in which he threatened that in the 1970s hundreds of millions of people would die of starvation, due to overpopulation and the limited capacity of the Earth to sustain their lives—therefore he adamantly called for birth control. Upon its publication, his book attracted quite a bit of criticism, and it was considered "pessimistic," especially in the late 1970s, since his predictions did not come true. Nonetheless, since then Ehrlich and his wife, biologist Anna Ehrlich (1933-), have continued to deal with issues of the Earth's limited resources and the question of the planet's ability to continue to sustain all the natural systems that support the existence of life, including that of Humanity. In later publications and interviews, the couple continued to warn that the lack of resources could lead to wars over resources, water and food crises, damage to biodiversity, and global warming, as well as global poisoning, due to excessive use of chemicals in agriculture, medicines, and more.

It is very difficult to estimate the maximum number of people the Earth can support. Moreover, many claim that, today, we are already beyond such an amount and, therefore, the natural resources are being irreparably depleting or the resources required are changing, while others claim that existing and future technologies will enable the Earth to support a higher larger population than the current one.

American biologist and naturalist, Prof. Edward Wilson (1929–2021), known as the "father of sociobiology" and of the concept of "species diversity," claimed in 2002, in his book *The Future of Life*, that if everyone agreed to be vegetarian, then the 14 billion dunams of agricultural land, available at that time, would be able to support ~10 billion people [15]. A U.N. report from 2012, reviewing many relevant studies, including various calculations regarding the Earth's carrying capacity, presented a very wide range of hypotheses, estimating from 2 to 1000 billion people, and concluding that the maximum number is 8 billion, very close to the current human population of Earth [16]. This report revealed that one of the main reasons for the differences between the various studies and the difficulty in estimating the maximum number of humans that the Earth could support is that there is no agreement on the estimation method and the reference scenario. A decade plus later, most of the estimates have converged at ~10 billion people.

The discussion regarding population growth and the Earth's carrying capacity for human society also raised many questions regarding planning and controlling births or limiting births. Over the years, birth limits have been established in several countries across the globe, starting in 1978 with the famous Chinese plan (the "one-child policy"), limiting the birth rate in China, until it was rescinded in 2015. Other such plans limited the number of children per family to two (in Iran, Vietnam, and Singapore). However, this issue is still quite controversial, due to conflicting socio-cultural and religious norms and values, besides issues of individual rights. Moreover, beyond the ethical and theological dilemmas, it raises concerns regarding the future existence of certain minority groups in the population, regarding the importance of maintaining diversity and variety within the human population, as well.

The Human Body

What and how much do we consume, and what and how much do we emit? The human body consists of many molecules and biological structures, each of which, small or large, plays an important role. It is customary to refer to the composition of the body by the five different levels of its structural components: the atoms, the molecules, the cells, the tissue-systems, and the body as a whole [17]. The most basic elements, the atoms, are the building blocks of every molecule. Of the 118 elements in the Periodic Table, the human body has ~50 elements, with 11 elements responsible for 99.5 wt%: oxygen (61 wt%), carbon (23 wt%), hydrogen (10 wt%), nitrogen (2.6 wt%), which constitute the organic matter and water producing 96 wt% of the body's weight, along with calcium (1.4 wt%), phosphorus (0.83 wt%), sulfur (0.20 wt %), potassium (0.20 wt %), sodium (0.14 wt %), chlorine (0.14 wt %), and magnesium (0.027 wt %), while the other elements are present in very small to minimal amounts [17]. In a general calculation for the body of an average person weighing 70 kg, this is ~6.9 * 10^{27} atoms.

The primary molecules that constitute body weight are: water (60 wt %, of which 34 wt % is water inside the cells and 26 wt % is water outside the cells); lipids—all the oil-soluble molecules (19.1 wt %, most of which are unnecessary fats—17 wt %, and the rest are essential fats—2.1 wt %); proteins—from amino acids to proteins in the cell nuclei (15 wt %); minerals—metals and nonmetals (5.3 wt %); and glycogen—a polysaccharide built from glucose (0.6 wt %).

An average person, weighing 70 kg, has ~10^{18} cells in his/her body. Although these cells share many similar characteristics, they differ from each other in size, shape, composition, and function. In general, it is customary to distinguish between four types of cell tissues: nerve tissue, epithelial (skin) tissue, muscle tissue, and connective tissue, which supports many other tissues. In addition to these tissues, there is "extracellular fluid" (ECF), ~94% of which is water, consisting of plasma (which constitutes ~55% of the composition of the blood and consists of 90% water plus proteins), as well as interstitial fluid, and various extracellular solids (organic collagen fibers from skin, bones, tendons, blood vessels, etc., and inorganics, like calcium and phosphorus, which constitute ~65% of the dry bone matter).

The level of tissues and systems refers to the functions of the various components. At this level, the muscles constitute 40 wt% of the body, fat tissue—21.6 wt%, the blood—7.9 wt%, the bones—7.1 wt%, the skin—3.7 wt%, the liver—2.6 wt%, the central nervous system (CNS)—2 wt%, the digestive system—1.7 wt%, the lungs—1.4 wt%, plus the remaining systems with negligible weights. Finally, the accepted, general anatomical division of the body is into head, neck, chest, back and abdomen, upper limbs, and lower limbs.

To sustain his body, a human consumes many natural resources, far more than his/her weight. This amount depends on many variables, including physiological variables, such as age, weight, sex, physical activity, even geographic variables (e.g., temperature and humidity), and sociocultural variables (like religion). Below are some key resources that every human being consumes.

Oxygen Oxygen is a vital element for life on Earth, including human life. An average person takes ~600 million breaths during his/her lifetime. The two lungs of an average adult contain ~6 L of air, and, on average, each inhalation draws on ~500 milliliters (mL) of air. The breathing rate of an adult is between 12 and 20 breaths per minute, 16 breaths on average, which totals 8000 mL or 8 L per minute. For the average life expectancy of ~82 years in the developed countries that comes to a total of ~345 million L of air, of which ~72 million L are oxygen.

Water Water is another vital substance for life, and it is the main component of the human body. Most of the water a person consumes comes from drinking and eating. Although people usually eat and drink every day, the human body loses water, mainly during urination, perspiration, and breathing. To maintain good health, the drinking of ~2–4 L of water a day is recommended, though the amount varies at different ages and for men and women. Yet, most of the direct water that humans consume is used for other purposes, such as bathing, cleaning, and irrigation. The average person consumes between 50 L per day in African countries and 400 L per day in the USA. The average domestic water consumption per day in developed countries is 120–200 L per person. Over the last 60 years, the total domestic consumption in the world has increased by 600%. The increased use of water in homes also leads to the production of much more domestic wastewater, 110–190 L per person per day in developed countries [18]. Moreover, pumping, conveying, and treating water for use are processes that require various chemicals and consume a lot of energy, all the more so if it is water desalination, which requires high-energy consumption during the salt removal process, while emitting air pollutants and GHGs. The production and treatment of wastewater also leads to the pollution of soil and water sources.

The "direct water" (i.e., water used at home, for drinking, bathing, cleaning, etc.) constitutes only ~8% of the water consumption of humans. All other water, called "hidden water" or "virtual water," is consumed indirectly, mainly in the water used for growing food, manufacturing clothes, and industrial products. For example, the water footprint of a kilogram of beef is ~15,500 L, compared to ~3400 L of water, which it takes to grow a kilogram of rice, and ~600 L of water to produce a liter of milk, when most of it is freshwater—"green" (rainwater) and "blue" (water in rivers, lakes, groundwater, glaciers, polar ice caps).

Not only food *drinks* water, but also many consumer products. The water footprint of a kilogram of cotton is ~10,000 L [19], while that of a pair of jeans is ~6300 L (which is equivalent to ~25,000 glasses of water)—with ~55% of the consumption coming from growing the cotton and ~40% from washing the pants themselves. Therefore, by means of a simple calculation, the amount of water required to produce a single pair of jeans is greater than the amount of laundry water used to wash it for all the years of its use. Even during the manufacture of an average family car, a large amount of water is used (~400,000 L for the production and processing of the metals, plastics, glass, and rubber)—an amount identical to the average family's consumption of direct water over 2 years. Similarly, the manufacture of a single mobile phone consumes ~1000 L of water during the production process, which is the average amount of water used per person, per shower, per month. Even the manufacturing process of a single sheet of paper consumes ~5–10 L, from the growth of a tree to the completion of the production process. As such, attempts to save water must include "source reduction" (i.e., reducing the use of direct water (freshwater sources), alongside the raising of awareness regarding the issues of "hidden water") and the wise consumption of food and textile products.

Food Measuring food consumption for a person ("per capita") or a country is much more complex, mainly because food consists of a variety of ingredients, the weights and nutritional values of which are different. It is customary to compare the consumption of peoples or countries based on the caloric values of their foods (i.e., the energy stored in what they eat), even if these values do not fully reflect the quality of that food. Daily food consumption per capita varies from 1700 to 3700 kilocalories in poor and developed countries, respectively. Furthermore, over the last 60 years, per capita food consumption in Asia and Africa has increased by ~55%, in North America by ~25%, and in Europe by ~17%. Accordingly, food production in the world is steadily growing.

The average person eats ~35 tons of food during his/her lifetime, but in developed countries, like the USA, it is twice that amount (~70 tons per person).

Consider the implications, according to various calculations, if an average person, who lived to the age of 80, ate during his/her lifetime: 4500 fish; 2400 chickens; 80 turkeys; 30 sheep; 27 pigs; and 11 cows. In developed countries, unfortunately "food waste" (i.e., food thrown away by the farmer, and/or the supplier, and/or the consumer) reaches ~33%, about two-thirds of it in the supply chain and about a third in the consumer's home. Moreover, the supply of food involves the utilization of many land resources (~90% of the agricultural land), water, and energy, as well as veterinary medicines, and chemicals for fertilization and pest control—that produce hazardous sewage and emit air pollutants and GHGs. Also mentioned above, many food and fiber crops are genetically modified.

The U.N.'s Food and Agriculture Organization (FAO) notes that today ~11% of the land on Earth is used to grow crops, and an even larger area is used to graze animals. The water footprint of food is very high. In fact, the water consumption of feed crops (animal food) constitutes ~27% of all the water consumption by Humanity. In addition, the consumption of food, especially what animals eat, has a significant effect on GHG emissions, especially methane (which is emitted by the digestion processes of sheep and cattle), and nitrous dioxide (which is emitted in the decomposition of fertilizer residues in the soil). Food production causes ~26% of the world's GHG emissions [20], while the food that is thrown away causes a quarter of those emissions (i.e., 6% of the annual GHG emissions in the world).

Energy According to the IEA, global energy production rose from ~7 terawatt-hours (TWh) in 1800 to ~160 TWh in 2020—a 2300-fold increase, but with the wastage of ~30% of the produced energy. The consumption of electricity per capita also varies dramatically from country to country, ranging from 0.1 to 51,700 kWh per year per capita. Finally, the global consumption of liquid fuels had reached ~100 million barrels (a barrel has 159 L of fuel) by the year 2020.

The "Culture of Consumption"

In a simple mass balance, the more the global population grows, the more natural resources it consumes for its existence and, as a result, those natural resources are diminishing, some even becoming depleted. However, the size of the world's population is not the only threat to the Earth and its resources—another serious threat is the growing per capita consumption that accompanies an increased quality of life. The population growth and consumer culture of human society also upsets the balance between the abiotic and biotic environments and between the different species in Nature, since more resources of land, water, food, metals, and more are diverted for the benefit of Humanity's existence.

Human activity also causes emissions of many pollutants, including the production of solid waste, liquid sewage, and emissions of hazardous gases into the air, and also emissions of GHGs. Many of the substances emitted during these processes are synthetic and "unfamiliar" to the natural environment, considered "pollution," as foreign substances located outside their normal purviews, where they should not be.

The culture of consumption involves many more products than are necessary for merely meeting the physical needs for the sustenance of Humanity (e.g., air, water, food, and energy); it also offers a wide variety of products and services that cater to and fulfill human desires.

Fashion The fashion industry is one of the largest manufacturing industries in the world. Every year between 100 and 150 billion new items are produced, 4 times the amount that was produced two decades ago. Theoretically, this would amount to 12–20 items per person, but private consumption differs markedly in different countries and societies. For example, the average acquisition for new items per capita in the USA is ~70 items per year, and ~50% of them items are discarded after a year. While at the end of 2018, the consumption of clothing and footwear in the world stood at 62 million tons, the expectation for 2030 is an increase of 63% (totaling 102 million tons), equivalent to the production of another 500 billion T-shirts! [21]. Furthermore, each year, clothing purchased at a cost of ~500 billion dollars is discarded without being used at all [22]. There is a usual "rule of thumb" (i.e., a commonly accepted approximation method) in the fashion industry—that ~20% of the clothes purchased will be used by the purchaser ~80% of the time. Furthermore, over the last decades, a "fast fashion" market has developed, referring not only to the short-term use of items, but also to the phenomenon in which market trends change at a rapid pace, with new designs and fashion details quickly reaching the public. This market is also characterized by speedy distribution, the increased purchase of cheap items, poorly sewn clothing made of low-quality fabrics, and items that wear out quickly after relatively little use. The fast fashion industry leads to the over-exploitation of perishable natural resources alongside modern slavery, mainly of women and children in backward countries, which makes it possible to sell disposable clothes at zero prices.

In developed countries, most items of clothing and footwear are used for a very short time and are discarded with all the material and energy resources that were used in their production. The mechanism of a "linear economy" treats fashion resources as perishable materials utilized in one-way processes taken from Nature, manufactured as commodities, purchased and used by consumers, and often discarded while still functional. However, recently, there is growing awareness of the negative environmental and climatic impact of the clothing and footwear sectors and their cloaked social inequality. This has spurred various attempts to expand the reuse and recycling of clothing items (e.g., some cities have bins for used clothing that is sold cheaply at "second-hand" or "vintage" clothing stores).

The fashion industry is one of the most polluting industries in the world. It uses ~215 trillion L of water and 168 million tons of chemicals to bleach and dye fabrics (about a quarter of the toxic chemicals in the world), as well as many fossil fuels, and is responsible for annual emissions of ~1.2 billion tons of GHGs (~2–8% of global carbon emissions) and other large amounts of various air pollutants, in addition to billions of tons of textile waste and about half a million tons of microplastics that find their way into the sea (~9% of the annual global amount discarded into the oceans) [23]. As mentioned before, the production of many fashion commodities has a very large water footprint, but also a very large carbon footprint. For example, throughout its entire life cycle, the production of one pair of jeans requires the use of an area measuring 12 m^2 (=14.28 yd^2) and it emits the equivalent of 33.4 kg of carbon dioxide. Another such example is the manufacturing process of leather clothes, which requires many natural resources (e.g., a lot of water), emits GHGs and wastewater (with many hazardous chemicals), and involves raising animals that may suffer. Thus, in recent years, there has been an extensive search for alternatives to leather clothing (beyond reduction at the source and the reuse of leather products). Some clothing has been made of plant materials and/or synthetic polymers. For instance, clothes made from "polyurethane" (an organic polymer), though produced from petroleum products and being non-degradable, is still a more eco-friendly option than natural leather.

Services The culture of consumption is also expressed by the constant rise in the utilization of various services. Unlike the production of physical products or tangible values, "service" is defined as the creation and delivery of intangible value, ranging from education and medicine to cultural and leisure services (based on mental resources like knowledge, expertise, time, etc.) It is true that these services are not directly connected to physical resources, but their provision does involve the indirect utilization of many natural resources manifested, for example, by traveling on airplanes or having water, linens, and towels while staying at a hotel.

The hospitality industry consists of four different sectors—lodging, food and beverage, leisure, and travel—and tourism was one of the fastest growing industries during the first decade of the twenty-first century, showing ~3% growth each year. The number of international flights increased from 25 million in 1950 to ~1.5 billion by the end of 2019. In addition, it is estimated that there are currently ~190 thousand hotels offering 17.5 million guest rooms worldwide. Other estimates point to ~400,000 hotels, B&Bs, guesthouses, and more, all without considering the market for rental apartments.

Not only does consumption vary from place to place, from time to time, and depends on a wide variety of economic and sociocultural variables, but it is also very difficult (almost impossible) to make a mass balance calculation on an individual's consumption, because it involves different resources, many molecules, and countless substances. One of the ways to deal with this issue, to enable the quantification of consumption culture, its management, and comparisons between people and between companies, is by measuring "consumption expenditures," both comprehensively and of each individual sector: food, energy, clothing, leisure culture, and so on.

According to various studies, the composition of consumption expenses by sector for EU households in 2019 was supply of electricity, water, and gas—23.5%; transportation—13.1%; food and non-alcoholic beverages—13.0%; recreation and culture—8.7%; restaurants and hotels—8.7%; furniture, home appliances, and household maintenance—5.5%; clothing and footwear—4.6%; health—4.4%; alcohol and tobacco—4.0%; communication—2.4%; education—0.9%; and the remaining services and other products—11.2% [24].

Another index proposed for the quantification of human activity and the consumption of natural resources is the "environmental footprint" or "ecological footprint." An "ecological footprint" is a calculation of the area needed to maintain the lifestyle of a population or a person [25]. This index incorporates indicators that measure the consumption of natural resources (i.e., the area needed to provide the food, energy, and products we consume, along with the area needed for living, leisure, and the treatment of the garbage and wastewater we produce). Thus, by calculating an ecological footprint, it is possible to compare the environmental impact of different countries. Opponents of this method claim that it does not quantify the socioeconomic factors that vary between countries and does not consider the quality of those processes. Indeed, the ecological footprint varies considerably between countries, ranging from 0.5 ha (5000 m^2 = 5950 yd^2) per capita in Eritrea in Africa to 16 ha (160,000 m^2 = 190,400 yd^2) per capita in Luxembourg, giving an average of 2.75 ha (27,500 m^2 = 32,725 yd^2) per capita [26].

Another way to quantify the human consumption of natural resources is to calculate the "overshoot day" (i.e., the day of the year when the human demand for resources exceeds the quantity of resources that the Earth can renew during that same year) for each country. In 2014, it had been determined that world's overshoot day was on December 19, while in 2021, it came on July 29 [27]—144 days (more than a third of a year) earlier. In 2025, the overshoot day for the USA was calculated as being March 13, for example. The consequences of ecological overshoot (the overconsumption of natural resources) is that Nature, once overdrawn, becomes unable to renew its own resources in a timely, self-sustaining manner, after which it can no longer provide essential natural resources nor sustain its own ecosystems.

Mobile Phones

Overconsumption is manifested regarding a wide variety of products, such as mobile phones. In 2023, ~1.14 billion smart phones were sold worldwide, compared to the 680 million sold in 2012. Moreover, in 2024, the number of smartphone users worldwide is 4.88 billion, meaning 60% of the world's population owns a smartphone.

A mobile phone consists of over 50 different elements of the 118 elements that exist in the Periodic Table, some of which (like various metals) are used in "pure substance" form, others within simple or giant molecules, or composite substances [28]. A typical mobile phone device weighs between 150 and 200 g. It contains: ~30 g of aluminum, ~20 g of iron, ~8 g of copper, ~6 g of cobalt (Co), ~5 g of chromium, ~3 g of nickel, ~0.34 g of silver, ~0.034 g of gold, ~0.015 g of palladium (Pd), and less than a thousandth of a gram of platinum (Pt). There are also many other elements in mobile phones: carbon, hydrogen, and oxygen in the plastic; silicon in the electrical panels and screen; and various metals (such as lithium) alongside rare metals (like thulium (Tm), lutetium (Lu), and ytterbium) in the battery. Moreover, the production of one mobile phone consumes ~1000 L of water, which is the average amount per person per shower per month. Moreover, the mining of these metals, the production of plastic, and the energy and radiation involved in the use of mobile phones, have far-reaching effects on both the natural and human environments.

Chemicals

The massive growth in world population motivated the building of cities, metropolitan areas, and megacities (i.e., urban areas populations exceeding 10 million people). In 1975, four such megacities were developing: New York, Tokyo, Mexico City, and São Paolo. Now, in 2024, there are many more (about 37) megacities, including Delhi, Shanghai, Dhaka, Cairo, Beijing, Mumbai, Osaka, etc., and others. The "undressing of consumerism" led to the overexploitation of natural resources and to the production and use of massive amounts of many chemicals [29]. In April 2021, the 250 millionth chemical was registered in the *Chemical Registry* of the Chemical Abstracts Service of the American Chemical Society. This *Registry* records elements, metals, minerals, acids, bases, organic substances, polymers, various mixtures, and more, gleaned from academic articles, patents, and registrations of businesses, companies, etc., classifying each chemical by means of a unique identification number (CAS RN). According to the CAS RN database, the growth rate of the registration of chemicals since the 1960s is exponential; just in the last decade, ~100 million chemicals were added to the database, ~40% of all the chemicals registered to date.

However, a 2020 comprehensive analysis of the industrial chemicals, listed in many national inventories, found a relatively small number—335,000—of registered chemicals [30], of which only 235,000 were with CAS RN, and the rest had some other form of registration or were listed as "trade secrets" (i.e., protected confidential intellectual property). Moreover, only 49,000 of those total chemicals are marketed chemicals (i.e., registered during the past decade), and ~261,000 chemicals were once traded or are still partially traded (unverified), while the rest are currently marked by a "preliminary registration" status (so it is not clear how many have actually been traded in the markets). In addition, of those 235,000 chemicals with CAS NR, 157,000 are pure substances of one kind or another, and the rest represent mixtures, polymers, and substances that cannot be precisely defined (e.g., substances that have unidentified ingredients, biological tissues, products of complex chemical reactions, etc.).

The amount of chemicals being produced in the world continues to grow. If, in 2017, the global mass of chemicals totaled 92 billion tons, by 2060, it is expected to stand at 190 billion tons. Furthermore, of all the manufacturing industries in the world, the chemical industry is the second largest. Since 1950, global production has increased 50-fold and is expected to triple in the coming decades.

The production and use of many chemicals in the sectors of agriculture, energy, transportation, infrastructure, and industry, and in countless products (e.g., foodstuffs, fashion, electricity, and others), cause the release of enormous amounts of chemical pollutants into the air, water, and soil, resulting in damage to plant, animal, and human health, as well as to habitats and ecosystems. Moreover, the rising rate of chemical production and the release of larger amounts of chemical byproducts and wastes (having diverse risk potentials) are overwhelming the present monitoring systems, such that insufficient, up-to-date data is available for the proper evaluation of their impacts on safety and health. The exposure of living creatures to airborne chemicals, just by breathing, and their ingestion, while eating and drinking, may cause morbidity or mortality. In the case of humans, there are many diseases triggered or caused by chemicals, ranging from cancer, respiratory and cardiovascular diseases, to allergies, reduced fertility, diabetes, dementia, and depression. In the case of the natural environment, chemical pollution causes the depletion of species diversity and disrupts or destroys ecosystem functions, essentially putting them out of service. In addition, massive chemical use is accelerating the climate crisis.

Pollution

"Pollution" is defined as the introduction of matter (solid, liquid, or gas, in forms that are: pure, mixtures, solutions, energy, heat, or radiation) into any environment in which they had not been before, or into a situation in which the rate of their addition to that environment is faster than the rate at which they can be harmlessly decomposed, diluted, or stored. Many substances are defined as "pollutants" having direct or indirect negative effects on the inanimate and living environments and/or on human health. It is customary to classify pollutants by substance type: chemical or biological, natural or synthetic, pure or in the form of a solution or mixture; and by their "states of accumulation": solid waste, liquid effluents, or gases. In addition, there are pollutants termed "primary" (i.e., emitted directly into the environment), and others termed "secondary" (created as results of a reactions by primary pollutants in the presence of additional matter and/or energy—which yields new substances and/or emits this or that energy). Moreover, it is also customary to differentiate between various pollutants according to the environment into which they are released: "air pollutants" emitted into the atmosphere, transforming from volatile organic compounds to particulate matter; "water pollutants" released into the hydrosphere, including oils, fuels, paints, and more; and "soil pollutants" released into the lithosphere, such as pesticides, fertilizers, and fuels. There are also other intangible forms of pollution, such as "noise pollution"—harmful noise that may cause negative health effects (e.g., headaches, deafness, or damage to the inner ear) or "light pollution"—the excessive or irregular use of artificial lighting, both indoors and outdoors. Indoors, excessively bright or glaring light may cause headaches, dry or sore eyes, and disrupt normal human sleep cycles; outdoors, glaring headlights may blind oncoming drivers, resulting in traffic accidents, and excessive lighting at night disrupts the natural patterns of local wildlife behavior, violating the natural balance in the species ecological habitat.

It is true that environmental pollution is often caused by natural events, ranging from volcanic eruptions to natural wildfires in forests, but most often, when we talk about "pollution," we are referring to the release of pollutants into the Earth's environment by Humanity, that is, from the anthropogenic source. This pollution—whether by the combustion of matter, or dumping sewage and waste, and whether by adding various substances to the soil, water, or atmosphere due to agricultural, domestic, and/or industrial processes—has existed since human society began to exist. Nonetheless, as global population continues to grow—and, with it, the use of natural resources and the production of a variety of materials and synthetic products—the quantity and types of pollutants are increasing, as well. Over time, this growth has also led to the development of various methods and measures for the control and treatment of pollutants, intended to prevent or, at least, reduce the release of toxic substances into our environment—at the source, via effective management, and by the proper treatment of hazardous wastes and sewage.

To reiterate, the sources of chemical pollution are many and varied. We engage them throughout our daily lives for most of our regular actions and they are involved in almost every product we use and service we enjoy. However, we now know that the generation of electricity in fossil-fueled power plants causes pollution of the land, air, and water. The alternative production of electricity in nuclear power plants (though somewhat more eco-friendly) produces radioactive wastes that must be treated accordingly, to prevent the emission of dangerous radioactive wastes. Our present industrial production, transportation, storage, and treatment of fuels—all these emit pollution into our atmosphere: particles, organic substances, oxides of sulfur, nitrogen, and carbon. A major cause of air pollution in populated urban areas is from the exhaust emitted by all the various means of transportation.

On the one hand, in rural, agricultural areas, soil fertilization and crop pest control have markedly increased crop yields per unit of land. On the other hand, the utilization of those chemicals pollutes the soil, air, and water sources, as do the sewage and waste byproducts associated with "animal husbandry" (i.e., the industry that raises animals for human food and animal food products)—which frequently contain, among other things, growth hormones, various other drugs given to the animals, and disinfectants—that may end up being unintentionally (and sometimes harmfully) ingested by human consumers.

The fashion sector, which manufactures processed, dyed, natural and synthetic fibers, is one of the sectors with the biggest environmental impact. This is partially due to the processing of synthetic fibers and the use of dyes produced from fossil fuel distillates—a process that requires many raw materials, much energy, and a great deal of water—as well as being due to the large quantities of problematic waste byproducts generated by discarded fashion products, along with plastic materials, metals, etc.

Plastic

"Plastic" is a general term for a variety of materials containing polymers, most of which are synthetic; "plastic" means "material that can be shaped" [31]. Different plastic materials are used in a wide variety of products, ranging from wrapping materials, shopping bags, and certain tools, to fibers used in clothing and footwear and some construction materials: pipes, furniture, vehicle parts, and more. Today various polymers are used in 3D printers that can print almost anything, from household items or clothes to "replacements" for certain body parts.

"Polymers" (from the Greek: *poly*—meaning "many" and *mer*—meaning "unit") are giant molecules consisting of repeated basic units called "monomers" (mono—one), linked together by chemical bonding. The "polymerization" process (i.e., the formation of polymers consisting of bonded monomers) may be done in several ways, usually by the addition or condensation of the monomers. In general, polymers may be divided into three major categories: "natural polymers" (or "biopolymers"), produced by plants or animals, such as cellulose; "synthetic polymers," artificially produced from compounds that usually originate in oil refining processes, such as polystyrene (PS) or polyvinyl chloride (PVC); and "semi-synthetic polymers," artificially produced from raw biological substances. It is also customary to distinguish between polymers produced from a single monomer (homopolymers) and those produced from two different monomers (copolymers). In addition to these, there are "linear polymers," whose "skeletal" chain structure does not contain large side groups; branched polymers, in which additional chains emerge from the skeletal chain; or cross-linked polymers, where different polymer chains form bonds between them. There are also two families of plastic materials: "thermosets" and "thermoplastics" [31]. The thermosetting polymers soften when heated, whereas after rehardening they cannot be softened again; they are usually composed of branched and cross-linked polymers. The thermoplastic polymers also soften when heated and harden when cooled, but they may be repeatedly heated and shaped in this way; they are usually composed of straight polymers. Table 5.1 illustrates some common synthetic polymers and their uses.

The uniqueness of the different plastic materials is that they are inert, relatively light, and relatively cheap and, above all, it is possible to design not only their structure but also their properties (e.g., for hardness and heat conductivity)—qualities that have led to its wide variety of applications and uses (Table 5.1). Nonetheless, the entire life cycle of synthetic plastic: from the extraction of the oil; the refining and separation of the monomers; the production of the various polymers, their transportation, and use, ultimately reaching the disposal of the plastic product at the end of its life cycle, has negative environmental effects. Unfortunately, since plastic waste does not decompose naturally (i.e., it is not decomposed into monomers by microorganisms), the large accumulations of plastic cause many negative consequences for the quality of the air, water, and soil sources that also affect biodiversity and Humanity.

Table 5.1 Common synthetic polymers

Polymer	Uses
Polyethylene	Packaging, bottles, pipes, and electrical wire coating
Polypropylene	Car parts, fibers, packaging, and wrapping materials
Polystyrene (PS)	Tableware and discs
Polyvinyl chloride (PVC)	Pipes, games, coats, flooring, and furniture parts
Teflon	Military products and cookware coatings
Nylon	Threads, ropes, and stockings

The massive use of plastic began ~70 years ago, and since then ~10 billion tons of plastic have been produced [32]. In 1950, there were 2.5 billion people in the world and 1.2 million tons of plastic were produced, while, in 2016, there were 7 billion people in the world and 320 million tons of plastic were produced—representing a 90-fold increase in the amount of plastic per capita. In fact, the forecast is that this amount will double by 2030 and triple by 2050! Every year, ~500 billion plastic bottles are produced in the world, and 4 billion disposable plastic cups are used every year in "Starbucks," currently the world's largest chain of coffee shops. Moreover, ~50% of the plastic in the world is produced for temporary use only (e.g., as disposable dishes or wrapping and packaging materials), so they have a very short life cycle, rapidly passing from manufacturing into the trash bin or becoming litter. So far, ~70% of the plastic produced on Earth has become waste matter, that is, ~7 billion tons of waste of which only 9% has been recycled and 12% was burned to recover energy, while the remaining was buried in landfills, garbage dumps, or in Nature.

Much of the discarded plastic ends up as litter on city streets and beaches, and a great deal of the waste plastic ends up in the oceans. Besides the aesthetic damage caused by plastic wastes, it directly harms animals that eat it or become entangled in it. Furthermore, plastic undergoes a process of weathering both under sunlight and solar radiation and under water, eventually crumbling into small particles (micro- and even nanoparticles). Such plastic particles end up almost everywhere: in our water sources and seawater, in the soil, in the air, in the arctic snow and rains, and, thus, in the food we eat and the water we drink. There are currently ~5.2 trillion pieces of macro-, micro-, and nanoplastic particles in the sea, reaching a total weight of ~75–199 million tons! [32] A recent study claims that every minute, plastic waste equivalent to the full load of a standard garbage truck (~20 cubic meters (m^3) = ~706 ft^3) of plastic is dumped into an ocean, reaching a global total of 11 million tons per year. This calculation predicts that by 2050, there will be more plastic in the sea than fish! [33] Moreover, in the Pacific Ocean, there is already a large plastic island containing ~1.8 trillion pieces of plastic.

Plastic is the dominant pollutant of marine environments, accounting for ~85% of the underwater pollution. The impact of plastic on the marine environment and its creatures is acute, often deadly, and includes potential entanglement, suffocation, starvation, drowning, bodily lacerations (external and internal), oxygen and light deprivation, and poisoning. A study found that plastic waste is found on Earth in 100% of the sea turtles, 60% of the whales, 40% of the water birds, and 35% of the seals, as well as in more than 100 thousand of the mammals and turtles in the sea, not including the million or so water birds that died of plastic pollution [34].

The plastic production process is also accompanied by large emissions of pollutants: from the outset of the extraction of the raw materials; during the oil refining process; and while producing the polymer. About 4% of the world's fossil fuels are currently used to produce plastic, and 99% of the world's plastic is synthetic (i.e., based on fuel distillates) [35]. Additionally, the production, use, and disposal of plastic causes large GHG emissions, and it is estimated that GHG emissions from the plastics sector will increase to the equivalent of ~2.1 gigatons of carbon dioxide per year by 2040 (more than India's annual emissions). Note that a certain portion of the plastic is burned illegally, while another portion is burnt under control in energy recovery facilities, but both cases emit many air pollutants during the combustion process, including carcinogenic substances.

To solve this menacing global plastic problem, several proposals have been put forward in recent years [36]. The cheapest, simplest, and most environmental solution is "reduction at the source" (i.e., reducing or stopping the use of plastic from the outset). The second solution is the recycling of plastic, as an end solution, after the waste products have already been created, used, and discarded. The third solution burns waste plastic to recover energy, while the fourth solution requires a transition to biodegradable bioplastic from renewable sources. "Reduction at the source" is based, first and foremost, on the principles of wise consumerism, on buying only necessary products that will be used over time, as well as making daily decisions based on environmental awareness of the ecological impact of the production, use, and disposal of various products. For example, switching from single-use plastic utensils and cutlery to reusable metal or glass ones. Source reduction may also be done when a "single-use" or reusable product is reused (e.g., reusing a (re)usable plastic container, bottle or dish repeatedly, for as long as it is still "usable" or buying second-hand plastic products). Note that reduction at the source may be done not only by the consumer, but also by the manufacturers, who can, for instance, design utensils with smaller amounts of plastic. In addition, collecting and recycling plastic or reusing recycled plastic to produce the same end-product should reduce the use of virgin plastic, while simultaneously diminishing the waste, filth, and pollution [37].

"Recycling" is a process in which the plastic is collected and separated into its various types and reprocessed as a raw material to create the same product or other products. In general, plastic can be mechanically recycled by shredding it or by chemically breaking it down into its basic components. Although this process reduces the amount of plastic produced and discarded, it is a process that requires high energy and water consumption and uses various chemicals—all of which affect the environment. Note that not all plastic products can be recycled; in many cases, a specific product consists of more than one sort of plastic, or in different colors, requiring a preparatory separation process. For example, plastic bottles made of polyethylene terephthalate (PET) can be recycled and turned into fibers, from which a brass jacket is made. However, thermosetting polymers, in which the chemical bonds are very strong, are not suitable for recycling (e.g., polycarbonate or PVC). Furthermore, during the recycle process, plastic may suffer degradation and lose some of its properties (e.g., transparency or strength). Therefore, in some cases, when plastic separation is complex or too expensive, or when the material has lost its properties or cannot be recycled, it can be burned by controlled incineration at energy recovery facilities, to recover its high energy content (despite the emission of air pollutants and GHGs).

Another alternative proposed in recent years is switching to "bioplastics" (e.g., vegetable fats and oils, corn starch, straw, woodchips, sawdust, and recycled food waste) from renewable natural sources—such as using cotton, instead of polyester, or choosing disposable dishes made of bamboo or corn [38]. Recall, however, that even the production of biopolymers requires resources (land, water, chemical fertilizers, and pesticides); they also pollute the environment, and often divert resources from the food industry to the plastics industry. The case of disposable utensils made of biodegradable plastic continues to perpetuate the culture of environmental consumption and wastefulness, while encouraging overconsumption; again, bear in mind that biodegradable utensils also need to be produced, packaged, transported, etc.—all of which have a heavy environmental price. Finally, it is important to realize that not all dishes marketed as "degradable" really are, and even if the dishes are truly degradable, they must be collected, separated, and transported to a composter, usually an industrial one, which is well-ventilated and treated properly, so that degradable organic waste is transformed into compost-type fertilizer, which can be returned to the soil.

Since plastic is a relatively new material, the complete information about its negative health effects is not yet available. However, the food-adjacent uses of plastic, whether as refrigerator or freezer storage containers, or microwave heating and cooking dishes, and as plates, cups, and cutlery, raise serious concerns regarding the potential leaching of harmful chemicals from the plastic into the food. Moreover, in recent years, a growing number of studies have shown that chemicals present in plastic or plastic particles that enter the body can affect human hormonal activity [39]. For example, the substance bisphenol A, first synthesized in 1891, is found in various "polycarbonates" (i.e., a group of thermoplastic polymers containing carbonate groups in their chemical structures) and used in a large variety of products: plastic baby bottles, food and beverage packaging, car parts, etc. These studies have shown that bisphenol A leaches from the plastic into the food and beverages. Already in the 1930s, it had been discovered that this substance acts like the hormone estrogen and, therefore, it may disturb the hormonal system. Furthermore, in animal studies, it was found that bisphenol A harms the animals' reproductive systems, thyroid glands, and liver functions, also causing veterinary heart disease. Additionally, "phthalates" (i.e., esters of phthalic acid), which are added to plastic to increase its flexibility and transparency, also found in many packaging materials, leak from food and beverage packaging. Phthalates are also widely used in the cosmetics industry, and some studies have found it to be a substance that disrupts human hormones and may damage fertility and thyroid activity.

Genetically Modified Food and Fiber

"Genetic engineering" is a scientific process that alters specific genes in the DNA of a certain organism to produce more desirable properties or to eradicate undesirable properties (such as the potential development of certain genetic diseases). Though humans have been enhancing the properties of some plants and animals for thousands of years, by means of selective breeding, gene selection, hybridization, etc.—serious DNA manipulation began only in the 1970s. Since then, many organisms, ranging from bacteria to plants and animals, have undergone "genetic modification" for educational, medical, agricultural, and industrial purposes.

Genetic engineering makes it possible to increase the resistance of organisms to pests, to enhance resistance to extreme environmental conditions, to extend longevity, to increase crop yields, to promote the production of specific components in cells, and to transform proteins into biopolymers or polysaccharides. Today, most of the food we eat is genetically engineered, as are the fibers used to make fabrics. For example, up to 92% of the corn, 94% of the soybeans, and 94% of the cotton in the USA are genetically modified [40]; and it is estimated that over 75% of the processed foods on supermarket shelves, ranging from spices and crackers to tomato sauce, contain genetically modified ingredients. Currently, there are many studies and even applied cloning processes in which DNA sequences, cells, tissues, and whole organisms are copied and reproduced—the most famous of which was the cloning of Dolly, the sheep, in 1996. Now, genetic engineering is also used for healing, by means of the introduction of genes, cells, or tissues, primarily in the treatment of hereditary diseases. In general, the goal of "gene therapy" is to replace a defective genetic sequence or a defective gene that causes a certain disease with a normal sequence or gene. This relatively new medical field is developing at quite a rapid pace [41].

Although genetic engineering has clear advantages, they come with many concerns about its impact on human health and the environment, as well as raising several social and ethical questions. For example, will plants or organisms that have been improved or produced by genetic engineering disrupt the natural balance by becoming dominant invasive species? [42]. In which cases does genetically modified food or medicine have a negative effect on the human body? [43]. Might genetic engineering produce currently unknown side-effects or harm plants, animals, and even humans (perhaps causing new allergies)?

The widespread application of genetic engineering also raises many ethical questions, such as: What should the limits of genetic modification be? What is the significance of enhanced longevity by means of genetically modified drugs or food? Does the human ability to do genetic engineering (as the establishment of human supremacy) significantly change the natural order—ultimately harming the other species and changing the ecosystems? [44]. Undoubtedly the most complex ethical issues relate to the processes of genetic engineering in humans. In recent years, a great revolution has been taking place in the field, thanks to the CRISPR system, a system derived from bacteria, in which the bacteria serve as a source of immunity against viruses by cutting the viral DNA. This makes it possible to edit genes, to excise certain genetic matter and replace it with another gene or genetic sequence, even in humans—to correct a genetic defect or alter certain characteristics [45]. There are also other novel technologies in the new field of "synthetic biology," in which man-made, synthetic molecules are used to produce behaviors known in natural biology. Even synthetic DNA is now being produced for a variety of uses [46].

The "Anthropocene Era"

As Humanity developed, we made less and less direct, local use of our nearby natural resources and environmental conditions, that is, we began to consume more and more products and services, produced from afar, thus indirectly consuming materials and energy. Simultaneously, the space in which most humans chose to live was becoming less natural and more synthetic or virtual. The man-made processes, for which Humanity is responsible, locally and globally, are seriously impacting natural processes.

A recently published study found that 2020 was the first year in history in which the quantity of synthetic inorganic mass produced by Humanity surpassed that of the total "organic mass" of all life on Earth [47]. This "inorganic mass" includes all the buildings and infrastructures, the plastic products, the various equipment and instruments, etc. It mainly consists of concrete, aggregates, bricks, and asphalt, alongside metals, glass, and plastic. According to this study, the mass of the buildings and infrastructures totals 1100 billion tons, surpassing the mass of all the flora in the world (900 billion tons). In addition, the global mass of all the plastic ($\sim$8 billion tons) comes to twice the total mass of all the animals on Earth. Furthermore, the demographic urban shift of the majority of the human population in addition to the higher quality of life and the ongoing production of synthetic materials have caused severe damage to natural environments. Thus, while the $\sim$8 billion people who exist on Earth today represent only $\sim$0.01 wt% of all living creatures since the dawn of Humanity, humans have caused the loss of 83% of the wild mammals, 80% of the marine mammals, 50% of the plants, and 15% of the fish, while the breeding of animals for food purposes has been tremendously increased [48].

There is a growing frequency of "revolutions" born of technological innovations and associated sociocultural changes, which are repeatedly causing chemical, geological, and biological changes in the matter of the Earth. In truth, Humanity, itself, is threatening its own natural habitat and its man-made human society. Might this suggest the inevitable arrival of a new evolutionary phase?

By the outset of the twenty-first century, the extensive depletion of natural resources and the flood of man-made chemicals pouring into Nature, accompanied by the severe loss of biological species diversity have instigated the determination of a new geological period, termed the "Anthropocene era" (i.e., the age of humans) [49], first coined in 2000 by Nobel Prize-winning Dutch climatologist and chemist, Prof. Paul Krutzen. This Anthropocene period reflects the fact that Humanity is shaping the geological structure of the Universe. It follows the previous Holocene period, which began at the end of the last ice age, 11,700 years ago, that had been characterized by the conquest of vast territories and accelerated population growth. Among the reasons given for establishing this new era are: the appearance of radioactive formations and various geological structures in recent decades (caused by human nuclear activity), the extensive use of synthetic materials (especially plastic), the creation of genetically modified materials, and increased emissions of GHGs and organic compounds into the atmosphere. Above all, this marks the huge changes that Humanity has instigated at the cost of natural chemodiversity and biodiversity.

The Ecological Crisis

The long series of changes in chemodiversity and biodiversity that have occurred on Earth severely effect the continued existence of the various natural processes, to the point of endangering the very existence of life on Earth and, particularly, the survival of Humanity. Furthermore, miniscule or massive changes in chemodiversity and biodiversity on Earth have led to far-reaching consequences in the atmosphere, lithosphere, and hydrosphere as well as in the biosphere and anthroposphere. Moreover, the man-made, synthetic substances and materials produced in the anthroposphere, eventually reach the natural environment, polluting it, often causing additional, harmful direct and indirect changes. As such, there are currently three major, intertwined, interactive, and ongoing *global* crises (Fig. 5.1):

1. The **chemical** or **pollution crisis**, referring to the emission of many pollutants into the natural environment, ranging from metals to organic materials, plastics of various types, etc.—all of which affect the diversity of species, damage habitats, and disrupt the functioning of ecosystems, ultimately harming human health.
2. The **biological** or **species diversity crisis**, referring to accelerated depletion of the natural species, due to the destruction or alteration of habitats, overexploitation of species, invasive species, chemical pollution, and climate change.
3. The **physical** or **climate crisis**, referring to the accelerated emissions of GHGs resulting from human activities and processed, thereby causing an ecological chain reaction, including the accumulation of GHGs in the atmosphere, the resulting greenhouse effect, and instigating accelerated global warming—as manifested by the constantly rising average temperatures on Earth and the creation of frequent and/or extreme climatic changes and events.

These three huge, simultaneous, global crises comprise the twenty-first-century's **ecological crisis**, in which the ecosystems on Earth, with their animate and inanimate components, are changing beyond recognition—many even failing, unable to maintain their functions, their millions of years of service to Nature and Humankind

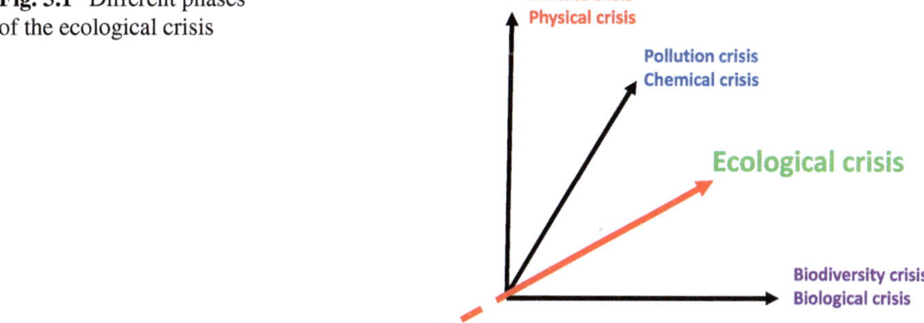

Fig. 5.1 Different phases of the ecological crisis

disrupted or completely disabled. Our "eco-social crises" reflect those negative impacts, not only on the vitality of Nature and its ability to sustain natural processes and renew resources, but also on human health issues, regarding the availability and/or accessibility of clean water, nutritional security, air quality, and the necessary development of preventive and/or therapeutic medical means for protecting vital systems in the human body from harmful man-made substances. The many services provided by Nature are essential for the existence of every species and ensure the continuity and prosperity of the ecosystems and of Nature as a whole. They may be divided into four general categories [50]:

1. **Provisioning services**—all tangible products produced by ecosystems, from raw materials and food to genes.
2. **Regulating services**—intangible services, such as temperature and humidity regulation, waste disposal, and disease control.
3. **Supporting services**—indirect services that enable the supply of products and the regulation of processes, such as the circulation of nutrients, pollination, and water and air purification.
4. **Cultural/spiritual services**—unlike the first three categories that refer to the natural processes of all the species, this category only applies to humans, referring to Nature as a major source of human inspiration and thought, one that provides spiritual and aesthetic experiences, as well as a place for respite and recreation.

To reiterate, the present ecological crisis was caused by human activity and is currently considered the greatest threat to the existence of life on Earth, threatening the safety and survival of Humanity itself. The growth of the world population and the rising standard of living have increased the degree of human interference in the natural ecosystems and initially manifested by the extraction of matter and energy from natural resources. The ensuing manifestation is the production of countless man-made substances, materials, and products (mostly synthetic) that are potentially harmful and incompatible with Nature's ecosystems. Then, when these man-made chemicals and synthetics (like plastics) are disposed of in Nature (as polluted wastes, sewage, and/or emissions), they inevitably do almost incomprehensible damage to our planet. Over the centuries, these human processes have caused far-reaching consequences in the types, quantities, and locations of matter and energy found in Nature, having markedly changed, disrupted, or destroyed the original natural ecosystems.

Planetary Boundaries

Human activity in the fields of agriculture, urbanization, and industrialization is the main factor affecting the natural environment today, having caused extensive and extreme changes in the Earth's natural systems and the current ecological crisis. Beyond the significant impact of the present ecological crisis on the natural environment, it also has far-reaching socioeconomic aspects. The following is an incomplete list of consequent socioeconomic damages incurred: loss of innocent human lives, outbreaks and spread of epidemics and pandemics, property loss, poor living conditions, conflicts over natural resources, population and refugee relocations, rising commodity prices, trade agreements failure, decease of economic performance and competitiveness of farms, inequality, etc. This crisis also has a long-term impact on the matter in Nature, ranging from elements and atoms, to molecules, compounds, and biological substances. All the man-made chemical, biological, and physical changes impact the natural recycling and restoration processes in the atmosphere, hydrosphere, lithosphere, and biosphere, creating internal imbalances within the natural processes and external imbalances (e.g., disruptions or disconnections) between the various resources and their reservoirs—thus, threatening the functional operations and existences of both natural and human systems.

In 2009, due to the realization of the great damage done to the balance of matter and energy in Nature, impacting both natural processes and Humanity, an international group of researchers from the fields of ecology, Earth sciences, and climatology proposed a new concept, known as "planetary boundaries." This methodology was further updated in 2015 and defined nine natural systems that support life, each of them with limits, within which Humankind might continue to exist safely on Earth [51]. They also quantified the amount of human activity that could be done within the predetermined limits, so Humanity would survive. According to the research, crossing these limits, or even one of them, would lead to radical and rapid changes in the environment, causing significant risk or catastrophe. These nine planetary boundaries contain the biogeochemical cycles in Nature (the nitrogen, phosphorus, carbon, and water cycles); most of the Earth's physical systems (the climate, atmosphere, and ocean systems); the Earth's biophysical characteristics that contribute to its basic robustness and self-regulatory capacity; measures of marine and terrestrial biodiversity; and two critical features related to anthropogenic global change—the "loading of aerosols" (i.e., solid or liquid particles suspended in the air we breathe, measured by their mass concentration) and chemical pollution.

The planetary boundaries proposed by the researchers are:

1. **Climate change**: A significant increase in temperature on Earth causes extensive climate changes. The defined limit is maintaining a concentration of carbon dioxide in the atmosphere of 350 ppm—current value (in 2024): 422.1 ppm.
2. **Ocean acidification**: The acidification of the oceans causes damage to marine biodiversity and disrupts the oceans' ability to absorb and successfully perform carbon dioxide fixation. An increase in the concentration of carbon dioxide in the atmosphere causes an increase in the acidity level of the water. The defined limit is a global average saturation state of calcium carbonate in surface seawater at a level of 2.75 omega (current value: 2.90 omega).
3. **Stratospheric ozone depletion**: The ozone in the stratosphere filters the UV radiation from the Sun's radiation. The defined limit is 276 Dobson units (DU) (current value: 283 DU).

4. **Biochemical flows**: The concentrations of nitrogen and phosphorus in marine systems have acute effects on the functioning of those systems and on the hydrosphere. Overfeeding marine systems with nitrogen and phosphorus can significantly damage aquatic ecosystems. The defined limits are 35 million tons per year of atmospheric nitrogen (current value: 121 million tons per year), and 11 million tons per year of phosphorus in the oceans (current value: 9 million tons per year).
5. **The erosion of biosphere integrity**: "Biodiversity" is a measure of the quality of ecosystems. The defined limit is an extinction rate of 10 species per year (current value: $\sim$100 species per year).
6. **Global freshwater use**: Today, humans have the largest impact on the global water supply. The limit defined for Humanity's use is 4000 km^3 of water per year (current value: 2600 km^3 per year).
7. **Land system change**: The expansion of agricultural activity alongside damage to forests demolishes habitats and destroys ecosystems. The defined limit is 15% of the Earth's land surface (not covered by ice) may be cultivated for agriculture (current value: $\sim$12%).
8. **Atmospheric aerosol loading:** A high concentration of aerosols in the atmosphere, particles dispersed in the air that serve as nuclei for the condensation of water, affect the climate crisis and public health. No limit has yet been set.
9. **Chemical pollution**: Human activity has led to the massive production of chemicals, many of them synthetic, causing serious damage to both the natural and human environments. To date, no limit has yet been defined; however, in a recently published article, it was stated that the increased production of chemicals and the different types of new chemicals have caused Humanity to exceed the planetary limit set for chemical pollution [52].

The Point of No Return

In recent years, a growing number of scientists have been talking about "climate tipping points" (climatic points marking highly significant, irreversible changes in the climate system). Such tipping points are expected to occur when certain temperature thresholds are exceeded and each one represents a different climatic change. In some cases, the consequences of crossing a certain climatic turning point may only appear hundreds or even thousands of years later, due to the complex and continuous chain of climatic processes.

A recently published comprehensive study found that we are already on the verge of five of the sixteen climatic tipping points examined [53]. Those five points are: the disintegration of the West Antarctic ice sheet; the disintegration of the Greenland ice sheet; the destruction of the coral reefs; the sudden thawing of the permafrost; and the collapse of heat conduction systems in the northern Arctic Ocean. As stated above, although the effects of passing any such tipping point may initially be local in the short term, ultimately, they will be global in the long term. Also note that certain changes are more visible than others; for instance, the melting of ice sheets and glaciers, due to the increase in atmospheric temperature, will cause a noticeable rise in the sea level when damaging habitats and ecosystems on land, as well as altering the wind regime, while also, sight unseen, changing certain ocean currents. Moreover, the demise of the coral reefs, due to the increase in ocean temperatures, will seriously damage marine biodiversity, because of the great reduction in the oxygen supply that had been readily available to the many marine species residing in those reef habitats. There, single-celled, microscopic plants, called "zooxanthellae," usually live inside coral polyps, engaging in oxygen-producing photosynthesis. Coral reefs produce a significant portion of the world's oxygen while covering only 0.0025% of the ocean floor. Besides containing oxygen, coral reefs also absorb carbon dioxide from burning fossil fuels. In addition, the coral reefs protect the beaches by moderating the strength of the waves; as such, once the coral reefs are gone, the beaches will likely deteriorate.

Crossing several climate tipping points could cause a more significant change in the entire climate system, leading to the point at which it fuels the changes and accelerates climate change on its own (i.e., having crossed "the point of no return"). The global climatic "point of no return," according to scientific estimates will be crossed in about a decade, if no significant action is taken to reduce harmful emissions. An example of such a process may be seen one day when large amounts of carbon dioxide and methane, currently trapped under frozen arctic areas, are released into the atmosphere after they thaw—no doubt they will accelerate global warming.

With the growing damage being done to the coral reefs and rainforests, it is becoming more and more difficult for Nature to successfully perform atmospheric carbon dioxide fixation by means of photosynthesis. Meanwhile, the rising temperatures are accelerating the anaerobic decomposition processes of organic matter in swamps and humid areas, giving rise to toxic methane emissions.

Another chain reaction that will cause further warming of the atmosphere, hydrosphere, and lithosphere is the "albedo effect." The albedo effect indicates the degree of deflection of the solar radiation bouncing off a surface, and the albedo of glaciers and snowy and frozen areas is very high. Once such highly reflective areas have melted, there will be much less radiation deflected into outer space and more heat will be absorbed into the ground, causing higher temperatures on the surface of the land and in the Earth's atmosphere. The albedo also gets smaller following deforestation and due to damage to forests caused by natural or man-made fires. Wildfires release large amounts of carbon dioxide and, afterward, the exposed ground, exposed to the sunlight, absorbs much more heat, having no protective green canopy. Eventually, the warming of the oceans and the temperature differential between the hydrosphere and the atmosphere may lead to a phenomenon in which the oceans will serve as hot reservoirs that gradually release heat into the atmosphere, heating the air in accordance with the principle of "thermal inertia" (i.e., the ability to store thermal energy received as heat, to preserve it, and release it gradually until it matches that of its surroundings).

Finally, it is important to note that the ecosystems may also play a part in the solutions to the various crises. Various natural ecosystems absorb the carbon dioxide that had been emitted into the atmosphere as a result of human activity, thus moderating the climate crisis. The albedo effect of glaciers cools the atmosphere, and trees absorb atmospheric heat, reducing local temperature. There are many other examples, such as rivers, given added flood control channels can safely channel flood waters, and certain microorganisms that decompose organic wastes and sewage may be intentionally utilized to reduce environmental pollution, rather than dumping untreated, hazardous pollutants into Nature.

References

1. Wolfson A (2016) Sustainable service. Business Expert Publisher, New York
2. Barker G (2009) The agricultural revolution in prehistory: why did foragers become farmers? Oxford University Press, Oxford
3. Wiersum KF (1995) Environ Manag 19:321–329
4. Shapin S (2018) The scientific revolution. University of Chicago Press, Chicago
5. Lucas RE (2002) Lectures on economic growth. Harvard University Press, London
6. Rifkin J (2008) Eng Technol 3:26–27
7. McChesney RW, Wood EM, Foster JB (1998) Capitalism and the information age: the political economy of the global communication revolution. Monthly Review Press, New York
8. Pathak P, Pal PR, Shrivastava M, Ora P (2019) Int J Eng Adv Technol 8:23–27
9. Rix J, Haas S, Teixeira J (eds) (1995) Virtual prototyping: virtual environments and the product design process. Springer
10. Shanahan M (2015) The technological singularity. MIT Press, Cambridge
11. Cordeiro JL (2021) Technological convergence: from biological evolution to technological evolution. In: Technological breakthroughs and future business opportunities in education, health, and outer space. IGI Global, Hershey, pp 23–41
12. https://ourworldindata.org/future-population-growth. Accessed Jul 2025)
13. Malthus TR (1798) An essay on the principle of population. Oxford University Press, London
14. Ehrlich PR (1968) The population bomb. Sierra Club, New York
15. Wilson EO (2002) The future of life. Vintage Books, New York
16. United Nations, Department of Economic and Social Affairs, Population Division (2013) World population prospects: The 2012 revision, Volume I: Comprehensive tables ST/ESA/ SER. A/336
17. Wang ZM, Pierson RN, Heymsfield SB (1992) Am J Clin Nutr 56:19–28
18. https://ourworldindata.org/water-use-stress. Accessed Jul 2025
19. Chapagain AK, Hoekstra AY, Savenije HH, Gautam R (2006) Ecol Econ 60:186–203
20. Poore J, Nemecek T (2018) Science 360:987–992
21. Global Fashion Agenda and The Boston Consulting Group (2018) Pulse of the fashion industry
22. MacArthur FE (2017) A new textiles economy: redesigning fashion's future. Ellen MacArthur Foundation, New York
23. https://www.unep.org/news-and-stories/press-release/un-alliance-sustainable-fashion-addresses-damage-fast-fashion. Accessed Jul 2025
24. https://ec.europa.eu/eurostat/databrowser/view/nama_10_co3_p3/default/table?lang=en. Accessed Jul 2025
25. Wackernagel M, Rees W (2004) The sustainable urban development reader. Routledge, London
26. https://worldpopulationreview.com/country-rankings/ecological-footprint-by-country. Accessed Jul 2025

27. https://www.overshootday.org/newsroom/past-earth-overshoot-days/. Accessed Jul 2025
28. https://www.compoundchem.com/2019advent/day16/. Accessed Jul 2025
29. Zaltman G, Zaltman L (2008) Marketing Metaphoria: what deep metaphors reveal about the minds of consumers. Harvard Business Press, Boston
30. Wang Z, Walker GW, Muir DC, Nagatani-Yoshida K (2020) Environ Sci Technol 54:2575–2584
31. Peters EN (2002) Plastics: thermoplastics, thermosets, and elastomers. Wiley, Hoboken
32. Geyer R, Jambeck JR, & Law KL (2017) Sci. Adv. 3:e1700782
33. Thushari GGN, Nadeeka G, Senevirathna JDM. https://www.pewtrusts.org/-/media/assets/2020/07/breakingtheplasticwave_report.pdf. Accessed Jul 2025
34. Thushari GGN, Senevirathna JDM (2020) Heliyon 6:e04709
35. Agenda I (2016) The new plastics economy rethinking the future of plastics. In: World economic forum, vol 36
36. Rohr JR, Martin LB (2012) Trends Ecol Evol 27:192–193
37. Hopewell J, Dvorak R, Kosior E (2009) Philos Trans R Soc B Biol Sci 364:2115–2126
38. Arikan EB, Ozsoy HD (2015) J Civ Eng Arch 9:188–192
39. Flaws J, Damdimopoulou P, Patisaul HB, Gore A, Raetzman L, Vandenberg LN (2020) Plastics, EDCs and health. Endocrine Society, Washington
40. https://www.ers.usda.gov/data-products/adoption-of-genetically-engineered-crops-in-the-us/recent-trends-in-ge-adoption.aspx. Accessed Jul 2025
41. Verma IM, Naldini L, Kafri T, Miyoshi H, Takahashi M, Blömer U et al (2000) Gene therapy: promises, problems and prospects. In: Genes and resistance to disease. Springer, pp 147–157
42. Sandler R (2020) Conserv Biol 34:378–385
43. Conner AJ, Jacobs JME (1999) Mutat Res Genet Toxicol Environ Mutagen 443:223–234
44. Bruce D, Bruce A (2014) Engineering genesis: ethics of genetic engineering in non-human species. Routledge, London
45. Pennisi E (2013) Science 341:833–836
46. Benner SA, Sismour MA (2005) Nat Rev Genet 6:533–543
47. Elhacham E, Ben-Uri L, Grozovski J, Bar-On YM, Milo R (2020) Nature 588:442–444
48. Bar-On YM, Phillips R, Milo R (2018) Proc Natl Acad Sci U S A 115:6506–6511
49. Crutzen PJ (2006) The "Anthropocene". In: Krafft T (ed) Earth system science in the anthropocene. Springer, Berlin, pp 13–18
50. Costanza R, d'Arge R, De Groot R, Farber S, Grasso M, Hannon B et al (1997) Nature 387:253–260
51. Rockström J, Steffen W, Noone K, Persson Å, Chapin III FS, Lambin E et al (2009) Ecol Soc 14:1–33
52. Persson L, Carney Almroth BM, Collins CD, Cornell S, De Wit CA, Diamond ML et al (2022) Environ Sci Technol 56:1510–1521
53. Armstrong McKay DI, Staal A, Abrams JF, Winkelmann R, Sakschewski B, Loriani S et al (2022) Science 377:eabn7950

Chapter 6
The Human Crisis

Abstract The fight against the human-driven ecological crisis requires immediate action on several fronts simultaneously. First, it is necessary to reduce the use of natural resources, production processes, pollutant emissions, and human population growth at the source. In addition, urgent efforts must be invested in research and development and in the implementation of various clean technologies in the energy, transportation, agriculture, and food sectors, as well as for transitioning to models of Green Chemistry and Green Engineering. The third front concerns changes in economic models—replacing capitalist models of growth that sanctify consumption with models focused on sustainability, prosperity, and greater equality, as does Doughnut Economics." However, without sociocultural alternatives and improved personal habits, no headway can be made against such a severe and impending crisis. It is, therefore, necessary to adopt models that integrate socioeconomic and environmental aspects (e.g., sustainability models) and to return to traditional values of cooperation, egalitarianism, and the promotion of the common good.

Keywords Green chemistry · Green engineering · Doughnut Economics · Sustainability

Thus far, Humanity has known many struggles between peoples: socioeconomic, cultural, religious, racial, and political, many local or regional, and two world wars, revolving around control over or damage to natural and/or other resources. Meanwhile, in routine daily life and in times of emergency, Humanity has further exacerbated the struggles between Humankind and the Earth. Such human conflicts, between human beings and between Humankind and the environment, have caused many changes in the natural environment, but have also caused the development of various socioeconomic models, meant to provide a "correct" and more effective response to the needs of Humanity and the distribution of resources among peoples and societies and with Nature. Nevertheless, in most of these models, humans are not viewed as being an integral part of Nature; rather Nature is considered more like a tool in the hands of Humanity, with almost no needs and rights of its own.

A. Wolfson, *Chemodiversity and the Ecological Crisis*, SpringerBriefs in Molecular Science, https://doi.org/10.1007/978-3-032-07363-1_6

Therefore, the various ongoing conflicts, alongside the political, socioeconomic, and cultural solutions they produced, have driven human-environmental relationship to negative tipping points, which ultimately brought about our current ecological crisis.

The word "crisis" represents a sharp turn, a decisive transition from one situation to another; but, more broadly, it is an unusual event or period with a negative impact on individuals, communities, companies, and/or organizations, and is often the result of a conflict that has taken a turn, one that does enable the involved parties to mediate or arbitrate successfully—a dead end. Unlike crises that arise from intrahuman conflicts, in the case of this ecological crisis (regarding species diversity; global pollution; the Earth's climate), it may seem that this is a natural event, divine fate, out of our control, although *we are the perpetrators*, each and every one of us. In fact, our present ecological crisis is the result of "a crisis of values" [1], which led Humanity not only to become disconnected and distant from the natural environment, but anthropocentric, self-serving, and hedonistic, considering only the here and now, and viewing personal needs, desires, and decisions narrowly and in the short term, without considering others or the natural environment. Although we talk about "the climate crisis"—it is not really a crisis of the climate—the climate did not create it, the climate is not responsible for it, nor can the climate resolve this problem. We are in the midst of "a human crisis," not in the sense of an individual human problem, or a social crisis, or even a humanitarian crisis—it is "a moral crisis"—the crisis that has shaped the Anthropocene era and caused our ecological crisis.

Modern culture places the "matter" in the center and pushes aside the "spirit." It sanctifies utilitarianism, which quantifies the value of each action according to its economic contribution and not according to its social and/or environmental value, nor by aspects of morality or social justice, encouraging Machiavellian individualism and an attitude of "Aprés moi, le deluge!" (meaning, "let the flood come once I'm gone," attributed to French King Louis XV, 1757). Such a culture, its socioeconomic and sociocultural models, and all its human interactions with Nature are based on the promotion of consumption and growth, processes that have greatly intensified since the Industrial Revolution, especially in this age of globalization, global society, and global markets. However, this cultural-economic process allows the continuous exploitation of nature and is responsible for most of the destruction of the natural environment. We should be amending our bad habits and addressing the ecological crisis by focusing on the reduction of natural resource depletion, seeking, and applying more natural alternatives that allow these precious resources to renew themselves.

In contemporary culture, Humanity is constantly looking for new and immediate thrills and gratification, perceiving technology and private property as primary tools for existence and self-determination. Nevertheless, while Humanity is constantly seeking to surpass limitations and boundaries by pursuing new technologies that maximize the quality of life, often in our own backyards, serious damage is being done to the quality of our environment!

However, make no mistake, the human crisis is not a crisis of the market economy or of market failure, nor is it a technological crisis, because technology is neither the problem, nor the solution. Our "human crisis" refers to day-to-day conduct, habits, and culture that have taken root over centuries, positioning humans as the dominant species. Now, the question is not only how to justly divide the remaining resources among people, but how Humanity and Nature can justly share them. The basis for dealing with the human crisis is, therefore, not found only in market corrections and not even only in the reduction of growth and consumption, that is, it is not merely a matter of changing economic or political structures, but rather of reinstating and updating the basic concepts and values of Humanity. The basic focus must be on morality, not on what is possible, so that we may strive for "prosperity," instead of "growth." We must learn to ask what is "appropriate" and not just what is "worthwhile" and to distinguish between what is truly needed and not what is merely wanted. A new social order must be established that puts Humanity back in its proper place—as part of Nature and one among many diverse species.

Since the climate crisis and its associated changes in chemodiversity and biodiversity are products of the human crisis, the only path towards reform is that of correcting misguided perceptions and habits, as well as redesigning harmful production, consumption, and growth assessment methods, while factoring in the functions and fates of atoms, molecules, substances, materials, products, and byproducts.

"Green Chemistry" and "Green Engineering"

The resources we did not consume are always the cheapest and most environmentally friendly. Therefore, many efforts are being made to reduce and improve the use of natural material and energy resources. An extensive effort is being made to find alternatives and new technologies that will make it possible to reduce consumption at the source and decrease polluting emissions and waste byproducts along the entire "value chain" (i.e., all the tangible and intangible value added to a product or service at all the stages of its development, from designer to customer). Some of these key solutions refer to a transition back to the use of natural raw materials, instead of synthetic ones and, in so far as is possible, to the production of chemicals and materials only from renewable and biodegradable sources. Recently, the energy sector has begun the process of transitioning from the combustion of fossil fuels to the use of cleaner, renewable energies; the search is still on for technological alternatives that will be able to meet human needs in a more intelligent and less polluting way.

For quite a while, industry was accused of being a major factor in the pollution of the natural and human environments and was targeted by environmentalists. This in addition to growing public awareness, regulation, and innovative (technologies have also led the industry to change its perception. Today, the relevant academic research deals with "upstream" solutions, consisting primarily of efficient and environmentally friendly, advanced planning processes, alongside "downstream" solutions for the reduction of environmentally polluting and GHG emissions. Such new models generally refer to the "life cycle" of a process or of a manufactured product, which starts with: the determination of its design and raw materials; material preparation processes; manufacturing processes; and the amount of material and energy resources to be utilized during its production, always bearing in mind the characterization of the desired product and its byproducts. Moreover, the new models also seek to examine the environmental impact of the final product, from the moment it leaves the factory until it has completed its life cycle (i.e., become nonfunctional or useless) and has been disposed of.

Yet, alongside the development of clean technologies, a change is also required at the atomic and molecular levels of the production of chemicals and other products. To that end, some new, alternative processes have been and are still being developed, the main goal of which is the planning and execution of more environmentally friendly processes regarding the molecules and atoms used throughout these entire processes. This endeavor, initially termed "green chemistry" in the 1990s, eventually became a distinct scientific field, later was also redesign as "green engineering."

"Green chemistry" is based on 12 principles developed in 1998 by two American chemists: Paul T. Anastas (1962-) and John Warner (1962-) [2], which may be adapted to any process:

1. **Prevention:** It is better to avoid the production of wastes, sewage, and atmospheric emissions than to clean and treat them after the manufacturing process. One of the main indicators used to examine the efficiency of industrial process emissions is the "E-factor calculations" (i.e., the ratio between the quantity of waste generated by a certain process and the quantity of its manufactured end product) [3]. This index deals, on a molecular level, with the wastes, excess reactants, byproducts, solvents, catalysts, etc., used during the reaction process that are not assimilated into the desired end products—that is, broadly assessing any and all the emissions resulting from an industrial process.

2. **Atom economy:** Chemical reactions and chemical or physical processes must be planned, while maximizing the combination of all the materials used during the process in the composition of the end products. There are two main indicators used to examine the efficiency of industrial processing. On the macroscopic level, it is calculated according to the "Environmental Performance Index" (EPI, which assesses the percentage of end products formed from the amount of the main reactant). On the microscopic level, it is calculated by "atom economy" (i.e., how many of the atoms found in the reactants are incorporated into the end products) [4]. Thus, the goal is to find alternative processes, in which other raw materials will often be used, which will yield maximum efficiency on the microscopic and macroscopic levels.

3. **Less hazardous chemical syntheses:** As much as possible, reactions and processes must be designed that use and produce non-toxic or low-toxic substances for human health and the environment. For instance, the production of chemi-

cals from carbohydrates, instead of hydrocarbons, and the production of biodegradable chemicals and products are preferable.

4. **Designing safer chemicals and products:** Chemicals and products must be designed that will maintain the effectiveness of their function over time, while reducing their toxicity and volatility.

5. **Use of chemicals and safe auxiliary materials:** Minimal use should be made of auxiliary chemical substances, such as solvents, salts, etc., and, only if necessary, the safest alternative substances should be used. In most processes, many substances are used during the reaction phase, and in the separation and cleaning processes at the end of the reaction. The use of these substances should be reduced and/or they should be replaced by less toxic and less dangerous substances.

6. **Design for energy efficiency:** The use of energy in the process should be reduced as much as possible. One should strive to carry out reactions at room temperatures and at atmospheric pressure.

7. **Use of renewable feedstocks:** The use of raw materials from renewable biological sources should be preferred whenever it is technically and economically feasible.

8. **Reducing the amounts of chemical derivatives in the process:** The production of unnecessary chemical derivatives must be minimized. Intermediates and byproducts requiring additional cleaning processes may create additional waste.

9. **Use of catalysis:** The use of catalysts enables more efficient, targeted reactions that produce a maximum of the desired end products and a minimum of byproducts. Catalysts enable substitution reactions to be carried out under milder conditions.

10. **Production of chemicals and products that decompose:** Chemicals and products should be designed so that, at the end of their function (life cycle), they will decompose into environmentally friendly byproducts that will not accumulate in Nature and may even be recycled into the natural reserves.

11. **Use of real-time monitoring and analysis to prevent the emission of pollutants:** The production process should include measures for real-time monitoring able to control the production and emission of polluting and dangerous substances.

12. **Plan production processes to prevent accidents:** Materials and processes must be carefully chosen to secure the minimal potential for accidents (e.g., the release of harmful materials into the environment, explosions, and fires).

Applications of green chemistry and green engineering are based on the desire to return to Nature and the use of renewable and biodegradable raw materials, utilizing most of the processing materials, thus preventing or reducing the emissions of byproducts, solid wastes, wastewaters, and air pollutants. In fact, green engineering processes, by saving raw materials and preventing needless pollution, are clearly acting to save the Earth's environment. Nonetheless, the prerequisites for the application of green engineering in industrial processes are altered perception and economic viability, especially since the environmental regulations are getting stricter and the external costs of such green processes keep increasing.

From Hydrocarbons to Carbohydrates

The fossil fuels, which Humanity uses as its main energy source today, are also the primary source of the raw material utilized in the chemical industry to produce a wide variety of man-made organic substances: ranging from chemicals for the plastics industry, and chemicals for medicines and cosmetics, to chemicals for the food industry. However, as previously stated, these fuels are not always biodegradable and emit large amounts of pollution during the processes of their extraction, refinement, production, transportation, and disposal, as well as producing various hazardous byproducts. In addition, many of the new synthetic molecules are also non-biodegradable and cause pollution when released into Nature; thus, extensive research has been and is being done to switch to the use of biomass as the primary raw material for the production of greener synthetic chemicals [5]. Moreover, various experts predict that in the coming decades fossil fuel feedstock will become very expensive, so that biomass-derived materials will be more competitive, not only from an environmental point-of-view, but also from an economic one.

The transition from chemistry based on fossil fuels to chemistry based on biomass is mainly a transition from chemistry based on hydrocarbons (originating from the oil refining process) to chemistry based on carbohydrates (i.e., natural sugars). For example, corn starch breaks down into lactic acid during a natural fermentation process in the presence of bacteria or fungi, which then reacts with ethanol to yield ethyl lactate ($CH_3CH(OH)CO_2CH_2CH_3$)—a substitute for toxic organic solvents. Another example is the use of natural color, flavor, and aroma substances extracted from plants and fruits, instead of those currently produced by means of synthetic processes from chemical compounds originating from the oil refining process.

Another growing field is that of "biological fuels" or "biofuels." "Biofuels" are fuels derived from biomass to serve as energy sources [6]. For example, "biodiesel" is derived from vegetable oil and fat, "bioethanol" is produced from sugar or cellulose, and "biogas," which contains mainly methane, is a product of the anaerobic digestion of biomass by microorganisms. Recently, the demand for biofuels is growing thanks to their environmental benefits. In fact, these fuels have existed for a long time. They were even used to power the first cars in the late 1880s and 1900s; for instance, in 1900, Rudolph Diesel's engine was running on peanut oil, and in 1903, Henry Ford's first car, the Model A, ran on ethanol. The outstanding advantage of biofuel sources is that, like mineral fuels, they provide energy for combustion processes do not require the complicated technological adaptation of existing systems. Moreover, their energy conversion efficiency is often similar to that of fossil fuels. Yet, unlike fossil fuels, the use of renewable biological sources and the ability to produce biomass fuel almost anywhere in the world could provide accessible and available energy for everyone everywhere. Although the conversion of biomass into energy is based on the process of combustion, which emits carbon dioxide—nonetheless, its use significantly reduces overall ecological damage and GHG emissions, because an equal or greater amount of carbon dioxide is assimilated by plant growth. In addition, it is significant that biofuels do not contain as many pollutants as metals, nitrogen, and sulfur compounds.

It is clear, however, that additional environmental aspects must be taken into account regarding the production of biomass, such as the proper uses of: land, water, fertilization, pest control, and more. There are issues regarding ethical trade-offs that must be considered, for example, only ~50 mL of bioethanol can be produced from the same amount of corn required to produce 1 kg of corn flour [7]. This being the case, growing corn for fuel production, instead of food production diverts resources from one market to another, increasing the price of corn flour, perhaps swaying the agricultural sector to repurpose its corn crop. Even the use of biofuels has a few bad environmental impacts, since shifting from gasoline to bioethanol increases the frequency of emissions that contribute to "eutrophication" (organic marine pollution) and photochemical ozone depletion.

Clean Technologies

Big and small revolutions throughout the history of Humanity all began with the appearance of new technologies that led to sociocultural changes. Now, when the climate crisis is signaling more strongly than ever, a growing number of technology companies are offering new "green" or "clean technologies" (or "clean-tech" in "high-tech," "bio-tech," "agro-tech," and "food-tech") able to reduce the exploitation of natural resources at the source and/or to pre-adjust solutions for the reduction of the emissions from industrial processes.

The battle against the ecological crisis has led to a race in search of wiser technological solutions for making cyclic use of natural resources, for streamlining man-made processes, and for the manufacture of sustainable products, in the fields of energy, transportation, water and wastewater, food, fashion, and more. Clean technologies are, first and foremost, technologies that place the emphasis on fewer resources being fed into each process and fewer pollutants being emitted by them. They rely on local resources and the transition of knowledge and skills from place to place, rather than the physical transfer of actual materials and energy. Their purpose is to mediate between development needs, the rising quality of human life, and the preservation of Nature. In general, it is customary to distinguish several areas dealing in clean-tech.

1. **Energy:** Searching for renewable, non-polluting sources of energy: solar, hydro-electric, and wind, as well as hydrogen and biofuel solutions (e.g., biodiesel and bioethanol), in lieu of fossil fuels. In addition, seeking to develop energy conservation systems and more energy-efficient processes.
2. **Transportation:** Focusing on improved mobility, rather than technological solutions based on private vehicles; development of mass transportation alongside electric and hydrogen-powered transportation; and the provision of walking and cycling routes.
3. **Water and wastewater (water-tech):** Development of more efficient systems for the utilization of water from existing reservoirs; for the capture and pooling

of rainfall; for the percolation and channelization of runoff; for water purification and wastewater treatment; and for desalination.

4. **Agriculture (agro-tech):** Development of automation, monitoring, and control systems, and especially "biological control" (i.e., the use of natural enemies instead of chemical substances), and the creation of alternative food sources (e.g., cultured meat produced from stem cells (**food-tech**).

5. **Prevention, reduction, and treatment of pollution**: Installation of monitoring devices, filters, and more, treatment of contaminated soils, and the development of clean chemical processes (**clean-tech**) and "green" chemicals.

6. **Climate**: Using innovative technologies to reduce carbon emissions and even absorb GHGs from the atmosphere, alongside technologies able to deal with and adapt to climate change (**climate-tech**).

These clean technologies have many advantages and provide diverse benefits. However, it is important to remember that today's world longs for and needs "breakthrough" technologies, those that do not suffice with merely attaining increased production at the expense of Nature's sustainability. We often tend to rely on this or that familiar technology, maintaining our harmful habits—thus, the existing technology may become part of the problem, rather than part of the solution. Now, the key questions are: Is advanced technology our tool or our goal? What kind of impact does each of our present technologies have on the natural and human environments? How can we restore a benign, sustainable balance between Nature and Humanity, before it is too late?

Renewable Energy

To meet the goals of air pollution and carbon emission reduction or zeroing, we must first actively change all our habits, in every area of our lives, while finding technological alternatives for our energy supply. Our goal must be to transition to the use of alternative and renewable energies (not based on carbon fuels)—a process known as the "decarbonization" of the "energy economy."

The sources of energy in the Universe may be divided into "degradable sources" and "renewable sources." "Degradable energy sources" are those whose utilization reduces their future usability, since their production rate is much smaller than their utilization rate. Such energy sources are mined from the Earth's crust and termed "mineral fuels." Among those currently serving Humanity as primary energy sources are "fossil fuels" originating from fossilized organic matter (animal or plant), such as: coal, petroleum, oil shale, and natural gas, and their combustion (as repeatedly stated above) emits air pollutants and carbon dioxide.

Another type of mineral fuel is "nuclear fuel," which can be used to produce "nuclear energy" either via "nuclear fission" (e.g., when a radioactive element, such as enriched uranium, is struck by neutrons and disintegrates into other substances, while releasing a large amount of energy), or via "nuclear fusion," in which two atoms fuse together, creating a heavier atom, while releasing a large amount of energy. The great advantage of these two processes is that they provide very high energy, many times more than is released by the combustion of fossil fuels and, most importantly, without emitting any GHGs. Nevertheless, while nuclear fission is currently being used to generate electricity, there is some risk of radiation leakage or a nuclear accident happening, which might seriously damage the surrounding environment. Additionally, all nuclear plants do produce "nuclear waste" that must be treated in a specific way, to prevent the radioactive contamination of its own environment. Nuclear fission, on contrary, is still much more expensive, due to the massive amount of energy required to trigger the reaction. Note, however, that many claim that, without the use of nuclear energy, it will not be possible to achieve the decarbonization of the energy sector and, as previously mentioned, today's technologies enable the use of smaller, more efficient, and safer reactors, thus reducing the risk of accidents and radioactive contamination.

The term "renewable energy" describes energy that may be replenished at a higher rate than it was consumed, such as solar energy; wind power; and water power; as well as the use of biomass (e.g., plant waste or other organic residues). Another accepted distinction is between energy that comes from burning organic matter, as opposed to "alternative energy" (i.e., energy unrelated to combustion processes), such as solar and wind energies. In recent years, another alternative has repeatedly arisen regarding fuel for heavy vehicles and power plants—the use of hydrogen, whose combustion reaction with oxygen produces harmless, eco-friendly water. In this case, the significant issue regarding its use is whether the hydrogen will be produced from water or fossil fuels. Finally, in addition to the use of renewable energies, it is also important to develop methods for energy conservation and energy efficiency, such as the proper maintenance of electrical systems and the insulation of systems and buildings, as wattage that was not produced is easier, cheaper, and the most viable, environmentally friendly alternative.

The competition between renewable and fossil energies is at the heart of today's climate and pollution crises. Many countries are increasing the share of wind, water, and solar energies in their mixed fuel array for electricity production and are promoting public and private electric transportation. Nevertheless, there are still technological gaps in the ability to store backup energy from renewable energy production during the "down hours" (i.e., when the process is inactive or not functioning). There is also an issue regarding the need for changing the electricity infrastructure in the wake of the decentralization of electricity production, due to the reliance on old fuels and old infrastructures.

Then, there is the matter of the prices and various geopolitical struggles, which repeatedly rattle the global energy economy. This means that the energy economy is subject to short-term and foreign interests, which highlights domestic and international socioeconomic disparities. Moreover, these mineral fuels are mined and extracted only in certain places, and are not found or available everywhere and, therefore, they become bargaining chips in local and global political conflicts of interest. Furthermore, there are environmental consequences from the production and use of fossil fuels for the climate, the quality of the human and natural environments, and public health, which the large and profitable companies have been hiding or diminishing all these years. In addition, there is no fair distribution of the good and the bad, and there is no social justice and distributive justice here.

First and foremost, switching to renewable energies that rely on free, available, natural resources allows everyone to harvest energy from Nature, anytime and anywhere. Furthermore, the production of energy for heating and/or electricity may be reserved for local use, avoiding the need for extensive transmission or storage infrastructures. Finally, the local production of energy makes energy consumers into energy producers, as well. Each individual or group initially produces enough to fill that individual's or group's needs and then sells the surplus energy to the network selling coverage to those suffering from shortages. In this way, there is enhanced personal or collective commitment to the smart consumption of energy by the producer-consumer (also termed "prosumer"). Additionally, the use of renewable energy makes it possible to promote "micro-grid systems"—autonomous emergency backup systems, connected to the main power grid, that produce the electricity they consume within a small, limited area. Together, all the above suggestions may produce an intelligent, more efficient, cost-effective, and cooperative electricity network, which combines smart and coordinated management of production and demand by means of advanced technologies for measurement, control, and communication between the consumers and the producers, and between different electrical products and systems.

Sustainable Transportation

The transportation sector is but one among the many sectors that require significant technological change. Transportation greatly affects the quality of the environment, in general, and the climate crisis, in particular. Transportation infrastructures require large areas, and various vehicles, powered by liquid fossil fuels, cause air pollution, GHG emissions, and noise hazards. The rather new concept of "sustainable transportation planning," which promotes the use of public transportation and encourages walking and cycling, is one of the many attempts to deal with the ecological crisis. However, the great dependence of modern society on convenient means of transportation, especially in private vehicles, has instigated the development of technologies that enable the use of alternative and renewable fuels (e.g., "electric vehicles" (EVs, with rechargeable electric batteries)) or "fuel cell electric vehicles" (FCEVs, with a tank of pure hydrogen converted to electricity by the fuel cell). As for the intelligent management of urban transportation, a growing number of applications have and are being developed to handle increasing use of public transport. Various "ride sharing" or "carpooling" services have also been developed, in which people ride together in a single car, thus reducing per capita pollution and carbon emissions. In addition, there are several navigation services that guide drivers in route, such as the "Waze" or "Moovit" applications.

"Cultured Meat"

Many moral, health, and environmental issues are associated with the raising, production, and eating of meat from animals. Recently, with the growing global population and the rising quality of life, on the one hand, and the climate crisis, on the other, efforts have increased for the commercial production of "cultured meat" (also known as "cultivated meat," "synthetic meat," or "lab-grown meat"; real meat created by the cultivation of animal muscle stem cells, without having the actual animal). Breeding animals to feed humans may cause the animals to suffer during their limited, unnatural lifetimes, in transportation, and when slaughtered. Animal meat, and especially red meat, is also a main source of saturated fat and cholesterol in the human body, often causing obesity. Moreover, red meat is associated with an increased risk of dying from cancer, heart attack, respiratory diseases, stroke, diabetes, various infections, Alzheimer's disease, and kidney (renal) and liver (hepatic) diseases. Furthermore, in animal husbandry, drugs are often used, ranging from antibiotics and growth hormones to steroids; these drugs are often passed on to the consumers through the meat they eat, which may also be a source of zoonotic diseases (as stated earlier) that may pass from animals to humans. Note, however, that eating animal meat also has certain health benefits, as a rich source of protein and iron.

Raising livestock for food also consumes many natural resources and emits pollutants into the air, soil, and water. One of the most detrimental effects of raising sheep and cattle is the increased emission of methane, released by the livestock's digestion and fermentation processes. In fact, the livestock on Earth are responsible for the emission of ~30% of the methane emitted by human processes globally—even more than the emissions released during the production and use of fossil fuels.

The advent of cultured or synthetic meat, grown under laboratory conditions, by means of the accelerated division of muscle stem cells in combination with plant protein, has recently been hailed as one of the most significant and promising solutions especially for the global reduction of methane emissions and for carbon emissions, in general, that may mitigate the climate crisis. Furthermore, switching from animal husbandry and the breeding of livestock for human consumption would greatly reduce the use of land and water resources; the amount of problematic sewage, waste byproducts, and polluting atmospheric emissions; as well as preventing some of the potentially negative effects of the meat on human health and the suffering of animals. Many research laboratories and bio-tech companies across the globe are currently racing to develop a cultured meat substitute that will be tasty, healthy, accessible, and cheap, and served in various forms, even a steak printed by a 3D printer!

Nonetheless, like any technology or alternative, the transition to "civilized" meat consumption also raises quite a few environmental questions, one of which deals with the energy issue. Currently, the production of cultured meat requires very high energy consumption; the electricity is still produced from fossil fuel; there are GHG emissions (even though most of them are carbon dioxide and not methane, they are higher than the carbon emissions from breeding livestock). Another issue is that of transportation of the end product, which also involves the emission of GHGs and many other pollutants. The issue of water is also significant; for the most part, the water is used to grow the source of the protein (e.g., soy) and increased use of water in the meat growing plant, which requires wastewater treatment and even recycling of the water. In addition, the issue of "overconsumption" (or "rebound effect") may arise, leading to food waste and more necessary waste treatment. Besides the race for cultured meat, there are currently technologies that offer protein alternatives to that found in beef or sheep, such as protein derived from soy, algae, or even grasshoppers.

Carbon Capture and Storage

Reducing carbon emissions refers to two main processes: reduction at the source during production processes and the consumption of food, clothing, electrical products, etc., and the transition to alternative energy. Yet, since it is impossible to reduce all the GHG emissions (not even those that are man-made) to reach the goal of a 100% reduction in carbon emissions (i.e., a state of "carbon neutrality" or "net-zero carbon"), it is necessary both to reduce carbon emissions at the source and to offset the byproducts emitted at the egress point (at the end of the process). Thus, it is important to implement proactive measures for the capture, fixation, and storage of carbon (i.e., carbon capture and storage) and, in cases of large quantities, to do so either prior to its emission into the atmosphere, or retroactively, by extracting it from the atmosphere. As mentioned above, the fixation of GHGs can be done naturally by trees and algae or by a variety of technologies for filtering, absorption, adsorption, and more. For example, it is possible to install absorption system in factory chimneys that absorb carbon dioxide in a solution with calcium ions, creating calcium carbonate. Similar processes occur naturally: when sea-shells and limestones are created; or carbon dioxide is adsorbed on various porous materials (i.e., adheres to their surfaces); or when the fixation of carbon dioxide in cement produces concrete, etc. Bear in mind that these technologies are currently in the early stages of their development and usage. For instance, there are already a few new technologies that make it possible to remove sparse GHGs from the atmosphere (despite their low concentrations); however, those technologies are still in their infancies and only work with very small outputs.

Finally, carbon offset and even achieving a "negative carbon balance" (i.e., with higher absorption and fixation than the amount emitted) are only possible in several scenarios. One such scenario is by using biofuels, during the production of which more carbon dioxide is assimilated into the plant than the carbon dioxide emitted during the burning of the fuel. Nevertheless, it is important to recall that the burning of fuels does have many other negative environmental effects: agricultural wastes must be properly treated; engaging in horticulture for biofuel production (e.g., biodiesel or bioethanol) consumes a great deal of land and water resources and promotes the inflation of the prices of goods and food. At this time, there are also premature technologies for extracting carbon dioxide from the atmosphere, even grandiose plans to extract the many emissions that have accumulated in the atmosphere since the beginning of the industrial age—but these very complex processes are still quite inefficient.

Sociocultural Alternatives

Although technology plays an important role in dealing with the ecological crisis, it is important to remember that technology is only the tool and not the goal, and that economic models can shape sociocultural changes. One of the key changes we must make is in the way we make decisions, especially in issues related to the utilization of natural resources. To deal with the many changes and the various manifestations of the present crises, it is important to have a dual, bifocal view of the local and global processes in natural and human systems—both a macroscopic view of large and small systems and the changes they undergo, and a microscopic view at the molecular level, to monitor detailed internal changes occurring within the systems (regarding the types, compositions, and variety of the components in the system).

Besides alternative processes and the use of clean-tech, concern over the phenomenon of environmental change and its consequences has led to global and local political efforts to deal with the crisis, both in terms of reducing emissions and in terms of adapting to changes and their consequences. Moreover, it was understood that, to bring about environmental change, it would be necessary to simultaneously respond to the social needs of humans (e.g., issues of social justice and well-being), while generating economic profit; this led to the formation and design of the "Sustainability Model." "Sustainability" is defined as the ability of Nature to maintain and support stable natural processes over time, while we take environmental, cultural, and socioeconomic aspects into account. There must be a balance between Nature's ability to continue to sustain the natural systems (ecosystems, life cycles, food chains, etc.) in ongoing support of life on Earth, and the development and needs of a certain population (human or otherwise) that depends on these processes. Simply put, we should strive to live prosperous lives today without diminishing the rights of future generations or depriving them of lives with conditions no worse than ours.

The progenitors of modern global sustainability models were initially developed in the early 1970s [8], and formed the bases for the various action U.N. plans regarding environmental issues, from "Agenda 21" (1992) to "Agenda 2030 for Sustainable Development," first presented in 2015. "Agenda 2030" offers Humanity a "to do list" that includes an action plan with 17 goals that must be implemented to lead to sustainable global development by 2030—Sustainable Development Goals (SDGs), and 169 targets. The founding principle of the 17 goals is to "leave no one behind," all with the aim of ensuring quality of life, human well-being, and prosperity for all and in a variety of fields.

The various goals composing "Agenda 2030" are directly or indirectly related to the type and quantity of natural resources in the natural environment, those used by Humanity, as well as the variety of synthetic substances and materials being emitted into the environment. Furthermore, the U.N. refers to the socioeconomic gaps that have been created between and within countries and societies, resulting from the unequal distribution of natural resources and different degrees of damage done to natural environments in various locations.

Economic Alternatives

Economic systems have a decisive influence on the consumption of natural resources and the emission of many substances into the natural environment. Therefore, the impact of the damage to Nature, when calculating the costs of economic processes, should lead to the more correct pricing of products and processes, thereby benefiting the natural environment. In economic terms, the ecological crisis is often defined as a product of "market failure," perhaps even the biggest market failure in human history. The concept of "a market" describes a relationship between "suppliers" or "sellers" and "buyers" or "customers," while a market economy refers to supply and demand interactions, which reach equilibrium and determine the price of a product or service voluntarily, and not coercively. A market failure occurs when the equilibrium becomes unbalanced, that is, an efficient allocation of resources does not occur and, therefore, an optimal result is not achieved. But the broader definition of "market failure" is a situation in which the "free market" (i.e., based on supply and demand without governmental regulation) does not maximize social welfare. In this context, an ecological crisis is defined as a market failure but, in fact, it is a combination of several distinct market failures.

The main reason for the market failure inherent in the ecological crisis lies in the fact that the impact of pollutant emissions, GHGs, and the loss of biological species on the market and social well-being are not quantified. These costs, called "external costs," and more precisely "negative external costs," that are indirectly involved in the production or consumption of a product, though not taken into account in the cost of the product itself—leading either to a deterioration in the production conditions of another product or to the decresed in standard of living of other consumer. More generally, it is a monetary value that expresses the loss of social welfare resulting from the emission of pollutants, GHGs, and damage to species' variety. Thus, when the external cost is added to the production cost, the "socioeconomic cost" is obtained. Most of these negative external costs will affect future generations, while those who caused the emissions and damage to the natural environment, its biodiversity, and chemodiversity will not pay the price. This is an intergenerational responsibility, based on intergenerational justice, one that refers to the excessive use of or harm to natural resources that will affect future populations, by inflicting restrictions and higher costs on their offspring.

Using external costs makes it possible to influence the reduction of emissions in different ways. One way is through enforcement and a demand for payment of damages for the emissions and their impact on the environment, whether as part of regulation (i.e., taxation and the like), or by means of supervision and enforcement, backed by legal action. It is likely that this will not stop most wealthy polluters from polluting, so pollution will continue. Although, once there is a price tag for pollution, many manufacturers may try to reduce polluting emissions at the source, to lower the cost of their damages, and reduce the risk of negative public relations or lawsuits. Note that this raises another socioeconomic quandary: Who should pay for polluting—the producer, the consumer, or both? For instance, does a company that produces electricity by burning polluting fuels have to pay the cost of the pollution or should it be the consumers enjoying the electricity, or both?

Another way to influence the economy is to use the external costs when making decisions and during planning processes and choosing between alternatives. Preference should be given to less polluting materials and processes, even if they cost more in terms of market costs. In this context, an extensive effort has been made all over the world, demanding that financial institutions (such as banks and insurance companies), and organizations and institutions (e.g., universities), divert their investments from fossil fuel companies.

Sustainable Economy Models

Today, in terms of natural resources, the global economy is based on two main principles—a linear use of resources, that is, take it from nature, use it, and throw it away, and "growth," which is defined as the development, production, and expansion of the possibilities in the economy to produce more products and increase consumption.

Thus, to deal with the shortcomings of these concepts, a different, cyclical economic paradigm is needed, one that imitates Nature, with material and energy resources moving in cyclical processes, termed a "circular economy" [9]. The circular economy is an approach that focuses on the conservation and regeneration of resources used in industrial systems and processes, to maintain their highest value for as long as possible, and to produce a minimum of waste. It uses a variety of strategies, among them the smart use of resources at the source, that is, in the production of the products, such as the use of perishable and renewable materials, on the one hand, and the reduction of the amount of raw materials, on the other hand, but also smart design, making it possible to reuse or recycle such products at the end of their "work lives." It is possible to extend the life of products and to reuse them as they are, or by repairing or renovating them (improving them in favor of the same use)—this in contrast to the currently prevailing, wasteful model of "design for failure," which reduces product lifetimes and usability to short periods. Efficient reuse and recycling have minimal negative impacts on the environment. In this way, a circular economy reduces the excessive use of raw materials and energy resources by redesigning them, so that they are "resource-poor," turning wastes into resources toward the production of new materials and products—often referred to as a "zero waste" economy.

The guiding principles of the circular economy are: (1) The smaller the cycle, in terms of economic activity and geographically, the more efficient and profitable it is; (2) Cycles have no beginning and no end, and the central principle is maintaining the value of the product without the addition of external value; (3) The speed with which the resources flow through the cycle is crucial; the efficiency of inventory management increases with decreased flow speed; (4) Continued ownership—reliance on reuse, repair, and renewal, without changes in the ownership of the product reduce environmental and economic costs; and (5) A circular economy must lead to change in the markets. In fact, the circular economy model seeks to promote closed-loop production patterns through the reuse and recycling of resources, and to optimize the use of resources to achieve a better balance between economy, environment, and society, while improving the conservation of natural resources and their renewal. Therefore, circular economy is changing not only the way resources flow through processes, but also the processes themselves. In this manner, the circular economy paradigm offers a restorative model, according to which renewable resources are used to reduce the environmental effects of the production of goods. Moreover, it proposes that every economic activity be planned so that the protection of Nature and its ecosystems is among its main goals. Furthermore, it offers a model with strategic and operational advantages, both at the micro and macro levels, which will fundamentally change the way we produce, consume, and use products, while changing our lifestyles. Finally, the vision of circular economy is to satisfy everyone's basic needs using limited amounts of natural resources, to move from a state of scarcity to a state of abundance.

In 1970, in the spirit of the circular economy and in imitation of natural processes, Swiss architect, Walter R. Stahel (1946-), while addressing the subjects on the extension of product shelf life and recycling of consumer products, coined the phrase "cradle to cradle," referring to the conservation of resources and the reduction of the amount of waste [10]. Later, in 1998, American architect/designer William McDonough (1951-) and German chemist Michael Braungart (1958-) published an article together, entitled: "The NEXT industrial revolution," in which they presented a new vision for sustainable planning and design, and a comprehensive approach for redesigning industrial processes [11]. Their approach dealt with the type and nature of the materials used, the degree of efficiency of the manufacturing processes, the use of those products over time, and the impact they would have on the environment. In 2002, these two authors presented a manifesto that challenged the concept of encouraging the "source reduction" or reuse and recycling of products—concepts that they both claimed—preserve the linear model of "cradle to grave" [12]. Instead, they called for integration into the natural cycles and producing products that would be used as biological nutrients, which could return to Nature and lead to prosperity, thereby significantly reducing the amount of waste, possibly even attaining a "zero waste" state [13]. This model also proposes considering the production process not only from an economic point of view, that is, in terms of the efficiency of material and energy use, but also examining how the use of the resources feed into the process and how those emitted from it affect the natural and human environments.

Alongside the model of the circular economy, old and updated economic models have developed, based on the wise use of resources and the reduction of damage to the natural environment, including "sharing" and "local" economies [14]. A "sharing economy" or "one-to-one economy" is an economic model based on borrowing, exchanging, or renting products or services, instead of purchasing them or owning them. This approach refers to a more circular use of natural resources, while increasing opportunities for individuals, communities, and companies, and serves as a social-community channel that allows everyone to enjoy the benefits of a certain product or service with minimal investment and damage to the environment.

The importance of local production and consumption are at the heart of "local economy," a model that links economy with community, and focuses on production and consumption activities within a certain geographic area, for the enhancement of personal and community prosperity, while preserving natural surroundings [14]. Local economy keeps resources and money within the state or community, while at the same time guaranteeing diverse and fair employment for the entire population.

Recently, another new and groundbreaking economic model was introduced, called the "Doughnut Economy," a model that began its development in 2012, but was only presented in 2017 by Kate Raworth (1970-), a British ecological economist [15]. Raworth's overarching motto is that the new economy should focus on prosperity, rather than growth, once access to the world's wealth (i.e., resources and knowledge) is fairly redistributed. Accordingly, the main challenge for Humanity is to meet the needs of each and every person within the limited resources and means available on Earth. According to this model, everyone must be provided with the essential elements of life: food, housing, education, health services, and more, in such a way that Humanity, as a whole, does not exert impossible pressure on the systems that support life on Earth. The "doughnut" that Raworth draws contains two central boundaries: internal social boundaries, which satisfy human needs, alongside external planetary boundaries, which represent the ecological ceiling that enables the continuation of life. In this manner, the doughnut economy seeks to balance the needs of human existence with the needs of ecological existence. Therefore, the uniqueness of this model lies in the fact that it offers a framework for sustainable development that connects economy, society, and environmental aspects.

Emission Economy

The "polluter pays" principle is an attempt to turn the external cost into an internal one, and to consider the connection between man-made changes in the natural environment and the damage they cause to Nature and Humanity; that is, to determine "price tags" for the emission of pollution, the growing production and emission of GHGs, and the harm done to species variety and ecosystems.

There is great difficulty in quantifying external costs, the price of resource depletion, or costs of shortages that may arise for those who come after us. Nevertheless, different methodologies have been developed for evaluating such external costs, most of which refer to damage caused by production, use, or disposal of products to human health, and sometimes also to environmental health (e.g., habitats, wild species, and ecosystems). Essentially, this is an attempt to set a price for pollution, which can be passed on to the producer and/or the consumer. To that end, the framework of "carbon economy," which focuses on reducing GHG emissions, can serve as an example of more chemically based solution. One of the economic tools currently use within this framework is called "emissions trading," in which each company has an annual emissions quota and valid emission permits. If a company emits fewer emissions than its valid quota, it can sell its remaining, unused valid emissions to another company, one that has exceeded its quota. International emissions trading also takes place between countries, in accordance with their specific emission caps. However, beyond the obligations of countries and companies, it is clear that a price tag on GHG emissions must also be imposed on consumers (i.e., the general public).

The primary goal of carbon economy is to transfer the responsibility and the burden of the damages to those who are responsible for it. One of the most discussed economic tools in this context is the "carbon tax." This tax is meant to add the price of the negative impact—that is, the external costs—of GHG emissions to the price of using fossil fuels, as well as to the price of energy and various consumer products. Thus, it is a "polluter pays" mechanism, which should ultimately bring about a reduction in consumption and use due to increased costs, thus eventually leading to a reduction in emissions. Moreover, while the "polluter pays" mechanism taxes the producers, and is often indirectly manifested by consumer prices, the "carbon tax" imposes a direct tax on consumers for promoting damage to the environment, in a mechanism popularly dubbed either "the user pays" or "the beneficiary pays." However, the use of a carbon tax also raises quite a few socioeconomic challenges, especially concerning the potential expansion of socioeconomic gaps. The main argument is not only that the weaker, disadvantaged sections of society are the first and main victims of climate change, whether due to damage to infrastructures or health, but that the imposition of any additional tax, of any kind, always hurts the less prosperous, those whose purchasing power is already very low. Yet, the degree of absurdity is even greater, since the higher the economic level, the higher the consumption of resources and products; as such, it is precisely the disadvantaged poor, who least contribute to global warming, who would be disproportionally and unjustly harmed by such a tax.

References

1. Nicholles N (2021) "A crisis of values" capitals coalition. https://capitalscoalition.org/a-crisis-of-values-by-natalie-nicholles/. Accessed Jul 2025
2. Anastas PT, Warner JC (1998) Green chemistry: theory and practice. Oxford University Press, London
3. Sheldon RA (2007) Green Chem 9:1273–1283
4. Trost BM (1991) Science 254:1471–1477
5. Lichtenthaler FW, Peters S (2004) C R Chimie 7:65–90
6. De Blasio C (2019) Fundamentals of biofuels engineering and technology. Springer
7. Yesmin MN, Azad MAK, Kamuruzzaman M, Ali S (2020) J Ecobiotechnol 12:1–4
8. Meadows DH, Meadows DL, Randers J, Behrens WW (1972) The limits to growth: a report from the CLUB OF ROME's project on the predicament of mankind. Potomac Associates Book, Washington
9. Liu L, Ramakrishna S (eds) (2021) An introduction to circular economy. Springer, Singapore
10. Stahel W (2016) Nature 531:435–438
11. McDonough W, Braungart M (1998) Atl Mon 282(4):82–92
12. Braungart M, McDonough W, Bollinger A (2007) J Clean Prod 15(13–14):1337–1348
13. Wolfson A (2016) Sustainable service. Business Expert Publisher, New York
14. Benington J (1986) Local Econ 1(1):7–24
15. Raworth K (2018) Doughnut economics: seven ways to think like a 21st century economist. Chelsea Green Publishing, White River Junction

Epilogue

The ecological crisis, in its various manifestations, is already here and now. It is an existing fact, on which there is a broad scientific consensus. We know who and what are causing this crisis and we understand its short- and long-term consequences for local and global environments. At this point, we even know a significant portion of the solutions that can and should be implemented at once, to prevent further exacerbation of this crisis and to mitigate its effects. However, despite the lively discourse on the subject (no longer contained within the research laboratories) and despite the academic and political discussions in mainstream media and public discourse—most of us are still acting as if nothing has happened.

The climate, pollution, and species diversity crises are results of the wasteful and predatory behavior of humans, and it seems that, as human knowledge developed and human technologies were perfected, the damage to the natural and human environments grew. Notwithstanding, the growth of the world's population and the constant increase in the quality of life divert more and more resources from the natural environment to the human environment, repeatedly violating the delicate balance required for the continuation and sustainment of the various natural processes. That is why it is so very important to maintain chemodiversity and biodiversity in Nature, and to rapidly revise the perceptions and amend the bad habits of Humanity.

In Nature, chemodiversity is crucial for the continued existence and functioning of natural processes, upon which human society depends. While there are substances in Nature, such as nitrogen and water, found in very high quantities, there are other substances, such as the metal palladium, the oxide of europium (Eu_2O_3), and hydrogen gas that are very rare. However, both the common and the rare materials in Nature are necessary for the existence of certain natural processes and systems. Biodiversity also plays a decisive role in the continued existence of natural processes and in the various services that ecosystems provide to Nature and Humanity. The quantity, type, and variety of species are measures of the wealth of Nature and of its ability to thrive, renew its resources, and survive.

A. Wolfson, *Chemodiversity and the Ecological Crisis*, SpringerBriefs in
Molecular Science, https://doi.org/10.1007/978-3-032-07363-1

We all know the importance of oxygen, nitrogen, or water, and even the roles of different metals (such as calcium and iron) in biosystems, but sometimes there are elements or compounds the biological or ecological values of which are not known to everyone, or not yet fully understood by science—even though there is no doubt that they are essential elements in the fabric of Nature. Moreover, excessive use of some of the natural elements or compounds in favor of human processes, whether in agriculture or in industry, also leads to a change in the availability of the various raw materials, especially for the creation of new materials. In Nature, elements and compounds appear in different states of aggregation, depending on the environmental conditions, and this variation also has great significance.

Modern humans have changed the Earth and Nature, for the worse, by producing synthetic materials unknown to Nature and unsuitable for its natural processes. Humanity has interfered in evolutionary processes, and developed many advanced technologies, to the extent that it has created a new historical period—"the Anthropocene era." However, the solution to our current ecological crisis cannot be based only on new technologies and requires updated and revised thinking and amended action (e.g., limiting the birth rate, switching to vegetarian nutrition, restricting vehicular traffic, etc.).

There is no doubt that we must rethink both the quantity and the quality of both resources and commodities, and manage material resources differently in the natural, human, and virtual environments. But while we tend to focus on what is visible to the human eye, the properties of natural and synthetic materials are related to standards that are invisible to humans. Thinking at the molecular level and planning economics at the atomic level will make it possible to maintain natural chemodiversity and biodiversity, while keeping Humankind in its rightful place and maintaining proper proportions—as one species among many in Nature, and as a body composed of diverse elements, molecules, substances, and materials.

These crises impacting the natural and human environments—the physical/climate crisis, the chemical/pollution crisis, and the biological/species diversity crisis—are all consequences of "the human crisis," stemming from a loss of values, as expressed in socioeconomic models, primarily in terms of increasing production, population growth, and rising over-consumption, and the exceeding of planetary boundaries, as well as a sense of human dominion over the Earth and estrangement from Nature. Increased food consumption, fast fashion, the habitual use of fossil fuels, and much more—in addition to hazardous atmospheric gas emissions and the dumping of polluted sewage and waste byproducts into the soil and water—have all contributed to global pollution. While Humanity is racing forward with all its might and on all fronts, it is no longer possible to wait. We must all act now, make personal changes individually, and collective changes together, if we want a favorable future.